T0142823

Low-overhead Communications in IoT Networks

Yuanming Shi • Jialin Dong • Jun Zhang

Low-overhead Communications in IoT Networks

Structured Signal Processing Approaches

Yuanming Shi
School of Information Science
and Technology
Shanghai Tech University
Shanghai, China

Jialin Dong
School of Information Science
and Technology
ShanghaiTech University
Shanghai, China

Jun Zhang
Department of Electronic & Information
Engineering
Hong Kong Polytechnic University
Kowloon, Hong Kong

ISBN 978-981-15-3872-8 ISBN 978-981-15-3870-4 (eBook)
https://doi.org/10.1007/978-981-15-3870-4

This Springer imprint is published by the registered company Springer Nature Singapore Pte Ltd.
The registered company address is: 152 Beach Road, #21-01/04 Gateway East, Singapore 189721, Singapore

Preface

The past decades have witnessed a revolution in wireless communications and networking, which has profoundly changed our daily life. Particularly, it has enabled various innovative Internet of Things (IoT) applications, e.g., smart city, healthcare, and autonomous driving and drones. The IoT architecture is established by the proliferation of low-cost and small-size mobile devices. With the explosion of IoT devices, a heavy burden is placed on the wireless access. A key characteristic of IoT data traffic is the sporadic pattern, i.e., only a portion of all the devices are active at a given time instant. In particular, in many IoT applications, devices are designed to be inactive most of the time to save energy and only be activated by external events. Thus, with massive IoT devices, it is of vital importance to manage their random access procedures, detect the active ones, and decode their data at the access point. Massive IoT connectivity has been regarded as one of the key performance requirements of 5G and beyond networks.

The emerging IoT applications have stringent demands on low-latency communications and typically transmit short packets containing both the metadata and payload. The metadata may include packet initiation and termination information, logical addresses, security and synchronization information, etc. It also contains a channel estimation sequence that facilitates channel estimation at the access point. Additionally, the metadata includes various information about the packet structure, e.g., the pilot sequences used for random access and device identification information. Considering the typical small payload size of IoT applications, it is of critical importance to reduce the size of the overhead message, e.g., identification information, pilot symbols for channel estimation, control data, etc. Such low-overhead communications also help to improve the energy efficiency of IoT devices. Recently, structured signal processing approaches have been introduced and developed to reduce the overheads for key design problems in IoT networks, such as channel estimation, device identification, and message decoding. By exploiting underlying system and problem structures, including sparsity and low rank structures, these methods can achieve significant performance gains. Chapter 1 provides more background on low-overhead communications in IoT networks and introduces general structured signal processing techniques.

This monograph shall provide an overview of four structured signal processing models, i.e., a sparse linear model, a blind demixing model, a sparse blind demixing model, and a shuffled linear regression model. Chapter 2 introduces a sparse linear model for joint activity detection and channel estimation in IoT networks with grant-free random access. A convex relaxation approach based on ℓ_p-norm minimization is firstly introduced, followed by a smoothed primal-dual first-order algorithm to solve it. For this convex relaxation approach, a trade-off between the computational cost and estimation accuracy is characterized by Proposition 2.1. The theoretical analysis of the convex relaxation approach is based on the conic integral geometry theory. This chapter only contains a brief introduction on the conic integral geometry theory. For more details, the interested reader can refer to Sect. 8.1 and other related mathematical literature enumerated in this monograph. Besides, an iterative threshold algorithm, namely approximate message passing (AMP), is introduced in Chap. 2, followed by the performance analysis based on the state evolution technique.

Blind demixing is introduced in Chap. 3, which facilitates joint data decoding and channel estimation without explicit pilot sequences. After presenting the basic convex relaxation approach for solving the blind demixing problem, we introduce three nonconvex approaches: the regularized Wirtinger flow, the regularization-free Wirtinger flow, and a Riemannian optimization algorithm. Theorems 3.1 and 3.2 provide the theoretical analysis of the convex relaxation approach and regularized Wirtinger flow, respectively. Furthermore, Theorem 3.3 presents the theoretical guarantees of the Wirtinger flow with the spectral initialization, which provides readers an easy access to well-round results. Readers who are interested in the intrinsic mechanism of the theoretical analysis can refer to Sect. 8.3 for more discussions. The theoretical analysis of the Wirtinger flow via random initialization is further provided in Sect. 8.4. Additionally, the basic concepts of Riemannian manifold optimization are presented in Sect. 8.5, which provide sufficient background for related algorithms in Chaps. 3 and 4. The extension of blind demixing, i.e., sparse blind demixing, is introduced in Chap. 4, which further takes device activity into consideration. The sparse blind demixing formulation is able to jointly consider device activity detection, data decoding, and channel estimation, for which three approaches are presented: a convex relaxation approach, a difference-of-convex-functions approach, and smoothed Riemannian optimization.

A further step to reduce the overhead is to remove the device identification information from the metadata. To support the joint data decoding and device identification, shuffled linear regression is introduced in Chap. 5. We first present maximum likelihood estimation (MLE) based approaches for solving the shuffled linear regression problem. Theorems 5.1 and 5.2 provide the statistical properties of the MLE, and both an upper bound and a lower bound on the probability of error of the permutation matrix estimator are introduced. To solve the MLE problem, two algorithms are presented: one is based on sorting, and the other algorithm returns an approximate solution to the MLE problem. Next, theoretical analysis of the shuffled linear regression problem based on the algebraic–geometric theory is presented. Based on the analysis, an algebraically initialized expectation–maximization algorithm is introduced to solve the shuffled linear regression problem, which enjoys better algorithmic performance than previous works. To give a comprehensive introduction of the algebraic–geometric theory, besides the concepts mentioned in this chapter, we introduce several related definitions on the algebraic–geometric theory in Sect. 8.7, including the geometric characterization of dimension, algebraic characterization of dimension, homogenization, and regular sequences.

Furthermore, Chap. 6 provides some cutting-edge learning augmented techniques for structured signal processing on the aspects of structured signal model design (e.g., structured signal processing under a generative prior) and algorithm design (e.g., deep-learning-based algorithm). We begin with compressed sensing under a generative prior, and other structure signal processing techniques under a generative model are worth further investigating, e.g., blind deconvolution. We next consider the joint design of measurement matrix and sparse support recovery for the sparse linear model (e.g., compressed sensing). Some basic methods are firstly presented, i.e., sample scheduling and sensing matrix optimization, and then learning augmented techniques are introduced. Additionally, for estimating the sparse linear model, several deep-learning-based AMP methods are introduced in this chapter: learned AMP, learned Vector-AMP, and learned ISTA for group row sparsity. In Chap. 7, we summarize the book and discuss some potential extensions of the area of interest. Tables 7.1 and 7.2 list the main theorems, propositions, and algorithms presented in this monograph.

The monograph is not only suitable for beginners in structured signal processing for applications in IoT networks but also helpful to experienced researchers who intend to work in-depth on the theoretical analysis of structured signals. For beginners, the background of both low-overhead communications and structured signal processing in Chap. 1 is helpful, and the problem formulation section in each chapter may be referred for further details with respect to each model. Tables 1.1, 7.1, and 7.2 provide quick references for the main results. Readers who are more interested in the intrinsic mechanism of the theoretical analysis of the specific models can refer to Chap. 8.

Low-overhead communications supported by structured signal processing approaches have received significant attention in recent years. The main motivation of this monograph is to provide an overview of the major structured signal processing models, along with their applications in low-overhead communications in IoT networks. Practical algorithms, via both convex and nonconvex optimization approaches, and theoretical analysis, using various mathematical tools, will be introduced. While the structured signal models concerned in this monograph have certain limitations, we hope the presented results will galvanize researchers into investigating this influential and intriguing area.

Acknowledgements

We express our gratitude to the support of National Nature Science Foundation of China under Grant 61601290. We also thank Xinyu Bian for proofreading an early version of the manuscript and Prof. Liang Liu for providing codes for part of the simulations in Sect. 2.4.

Shanghai, China — Yuanming Shi
Shanghai, China — Jialin Dong
Kowloon, Hong Kong — Jun Zhang

Contents

Mathematical Notations

- The set of real numbers is denoted by $\mathbb{R}$ and the set of positive real numbers is denoted by $\mathbb{R}_+$. The set of complex numbers is denoted by $\mathbb{C}$. Denote $\mathbb{S}_+$ as the set of Hermitian positive semidefinite matrices. Moreover, $\mathbb{N}$ represents the set of natural numbers.
- The boldface and lowercase alphabets, e.g., $\boldsymbol{x}$, $\boldsymbol{y}$, denote vectors. The zero vector is denoted by $\mathbf{0}$. A vector $\boldsymbol{x} \in \mathbb{R}^d$ is in the column format. The transpose of a vector is denoted by $\boldsymbol{x}^\top$. The complex conjugate of $\boldsymbol{x}$ is represented as $\bar{\boldsymbol{x}}$. The conjugate transpose of a vector is denoted by $\boldsymbol{x}^\mathsf{H}$ or $\boldsymbol{x}^*$. x_i denotes the i-th coordinate of a vector $\boldsymbol{x}$.
- For a complex vector $\boldsymbol{x}$ or a complex matrix $\boldsymbol{X}$, the real parts of them are represented by $\Re\{\boldsymbol{x}\}$ and $\Re\{\boldsymbol{X}\}$, respectively. Likewise, the imaginary parts are denoted as $\Im\{\boldsymbol{x}\}$ and $\Im\{\boldsymbol{X}\}$.
- The boldface and uppercase alphabets, e.g., $\boldsymbol{A}$, $\boldsymbol{B}$, denote matrices. A_{ij} denotes the element at the i-th row and the j-th column.
- The support function of a vector $\boldsymbol{x}$ is denoted as

$$\mathrm{supp}(\boldsymbol{x}) := \{i : x_i \neq 0\}.$$

A vector $\boldsymbol{x}$ such that $|\mathrm{supp}(\boldsymbol{x})| \leq s$ is defined as s-sparse.
- For a vector $\boldsymbol{x} \in \mathbb{R}^d$ or $\boldsymbol{x} \in \mathbb{C}^d$, its ℓ_p-norm is given by

$$\|\boldsymbol{x}\|_p = \sum_{i=1}^{d} |x_i|^p.$$

In certain cases, we define ℓ_0-norm as $\|\boldsymbol{x}\|_0 := |\mathrm{supp}(\boldsymbol{x})|$.
- For a matrix $\boldsymbol{A} \in \mathbb{R}^{m\times n}$ or $\boldsymbol{A} \in \mathbb{C}^{m\times n}$, the Frobenius norm of $\boldsymbol{A}$ is defined as

$$\|\boldsymbol{A}\|_F := \sqrt{\sum_{i,j} |\boldsymbol{A}_{ij}|^2} = \sqrt{\sum_i \sigma_i(\boldsymbol{A})^2},$$

where $\sigma_1(A) \geq \sigma_2(A) \geq \ldots \geq \sigma_{\min\{m,n\}}(A)$ denote its singular values. The nuclear norm of $\boldsymbol{A}$ is denoted as $\|\boldsymbol{A}\|_* := \sum_i \sigma_i(\boldsymbol{A})$. The spectral norm of a matrix $\boldsymbol{A}$ is denoted as

$$\|\boldsymbol{A}\| := \max_i \sigma_i(\boldsymbol{A}).$$

- The cardinality of a set $\mathscr{S}$ is denoted by $|\mathscr{S}|$.
- Random variables or events are denoted as uppercase letters, i.e., X, Y, E.
- The indicator function of an event E is denoted by $y = \mathbb{I}(E)$, where $y = 1$ if the event E is true, otherwise $y = 0$.
- Throughout this book, $f(n) = \mathcal{O}(g(n))$ or $f(n) \lesssim g(n)$ denotes that there exists a constant $c > 0$ such that $|f(n)| \leq c|g(n)|$, whereas $f(n) = \Omega(g(n))$ or $f(n) \gtrsim g(n)$ means that there exists a constant $c > 0$ such that $|f(n)| \geq c|g(n)|$. $f(n) \gg g(n)$ denotes that there exists some sufficiently large constant $c > 0$ such that $|f(n)| \geq c|g(n)|$. In addition, the notation $f(n) \asymp g(n)$ means that there exist constants $c_1, c_2 > 0$ such that $c_1|g(n)| \leq |f(n)| \leq c_2|g(n)|$.
- For a general cone $C \subset \mathbb{R}^d$, the *polar cone* C° is the set of outward normals of C:

$$C^\circ := \left\{\boldsymbol{u} \in \mathbb{R}^d : \langle \boldsymbol{u}, \boldsymbol{x} \rangle \leq 0 \quad \text{for all } \boldsymbol{x} \in C\right\}.$$

The polar cone C° is always closed and convex.

Chapter 1
Introduction

Abstract This chapter presents a background on low-overhead communications in IoT networks and structured signal processing. It starts with introducing three key techniques for low-overhead communications: grant-free random access, pilot-free communications, and identification-free communications. Then different models for structured signal processing to support low-overhead communications are presented, which form the main theme of this monograph. A classical exemplary of structure signal processing, i.e., compressed sensing, is provided to illustrate the main principles of algorithm design and theoretical analysis. Finally, the outline of the monograph is presented.

1.1 Low-Overhead Communications in IoT Networks

The proliferation of low-cost and small-size computing devices endowed with communication and sensing capabilities is paving the way for the era of IoT. These devices can support various innovative applications, including smart city, healthcare [1], and autonomous driving [22] and drones [27]. The explosion of IoT devices places a heavy burden on the wireless network, as they demand scalable wireless access, which has been put forward as a key challenge of 5G and beyond networks [23]. A key characteristic of IoT data traffic is the sporadic pattern where only a small part of all devices are active at any time instant. In particular, in many IoT applications, devices are designed to be inactive most of the time to save energy and only be activated by external events [26]. Thus, with massive IoT devices, it is of vital importance to manage their random access procedure, detect the active ones, and decode their data at the access point.

Moreover, the emerging IoT applications have stringent demands on low-latency communications, and typically transmit short packets containing both the metadata and payload [16]. An exemplary packet structure is illustrated in Fig. 1.1 (please refer to [29] for more details). The metadata may include packet initiation and termination information, logical addresses, security and synchronization information, etc. [16]. In the example showed in Fig. 1.1, we simply illustrate the metadata that contains a preamble and a header coming from the media-access-control

Y. Shi et al., *Low-overhead Communications in IoT Networks*,
https://doi.org/10.1007/978-981-15-3870-4_1

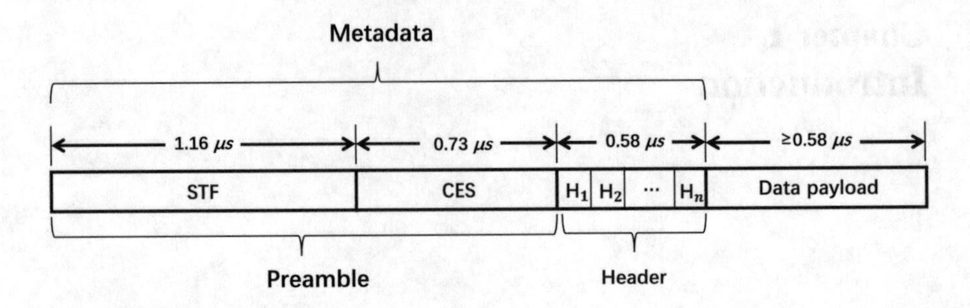

- STF: short training field
- CES: channel estimation sequence
- H_1 : pilot sequence
- H_2 : device identification information

Fig. 1.1 An exemplary packet structure

(MAC) layer and physical (PHY) layer. Specifically, the preamble contains a short training field (STF), which will be used for packet detection, indication of the modulation type, frequency offset estimation, synchronization, etc. It also contains a channel estimation sequence (CES) that facilitates channel estimation at the access point. Additionally, the header includes various information about the packet structure, e.g., the pilot sequences used for random access and device identification information. The header also includes the modulation and coding scheme adopted for transmitting the data payload. Furthermore, it may include the length of the payload and a header checksum field [16].

From the packet structure in Fig. 1.1, we see that the efficiency of short-packet transmissions, in terms of energy, latency, and bandwidth cost, critically depends on the size of the metadata, which is comparable to the payload size in many cases. To improve the communication efficiency, plenty of efforts have been made to reduce the size of the metadata, which result in *low-overhead communications*. Reducing overheads will not only improve spectral efficiency, reduce latency, but also achieve significant energy saving, which is especially important for resource-constrained IoT devices. In the sequel, we shall introduce three representative methods for reducing overheads.

1.1.1 Grant-Free Random Access

Conventionally, the grant-based random access scheme (illustrated in Fig. 1.2a) is applied to allow multiple users to access the network, e.g., in 4G LTE networks [3, 19, 26]. Under this scheme, each active device randomly chooses a pilot sequence from a predefined set of orthogonal preamble sequences to inform the base station (BS) of the device's active state. A connection between the BS and the active device

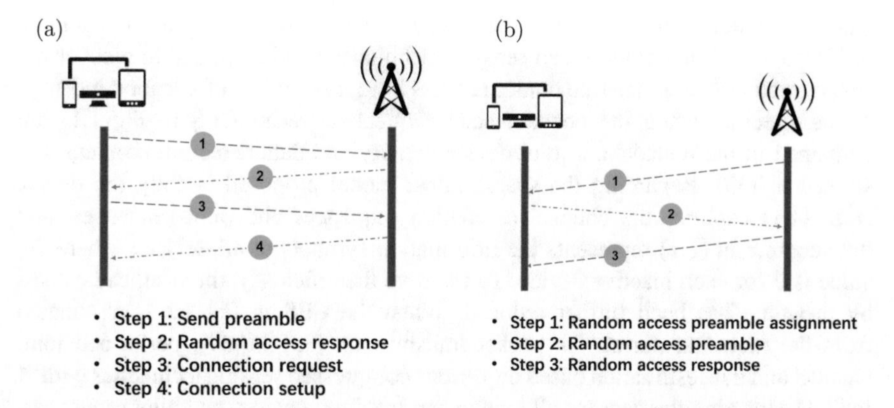

Fig. 1.2 Random access schemes. Note that for the grant-based scheme, steps 1–3 may need to repeat multiple times to establish a connection due to contention. (**a**) Grant-based. (**b**) Grant-free

will be established if the pilot sequence of this active device is not occupied by others. In this case, the BS will send a contention-resolution message to inform the device of the radio resources reserved for its data transmission. If two or more devices have selected the same pilot sequence, their connection requests collide. Once the BS detects this collision, it will not reply with a contention-resolution message. Instead, the affected devices have to restart the random access procedure again, which leads to high latency. Note that the messages sent by the active devices in the first and third phases correspond to metadata, as they are control information for establishing the connection without carrying any payload. Besides the overhead, a major drawback of the grant-based random access scheme is the limited number of active devices that can receive the grant to access the network. For example, as shown in [26], for a network with one BS and 2000 devices, a minimum length of the pilot sequence of 470, out of the total 1000 symbols, is needed to guarantee a 90% success rate. Even equipped with advanced contention-resolution strategies [6], 930 out of 1000 symbols are still required for transmitting the pilot sequence.

To address the collision issue of the random access scheme caused by a massive number of devices, the grant-free random access scheme illustrated in Fig. 1.2b has been proposed. With this new scheme, the devices do not need to wait for any grant to access the network and can directly transmit the coupled metadata and data to the BS. In this way, the BS can perform user activity detection, channel estimation, and/or data decoding simultaneously [33, 36–39]. The essential idea underlying this line of studies is to connect with sparse signal processing and leverage the compressed sensing technique. In particular, a compressed sensing problem is established by exploiting the sparsity in the user activity pattern. The received signal at the BS equipped with a single antenna is given by a *sparse linear model*:

$$\boldsymbol{y} = \boldsymbol{A}\boldsymbol{x}, \tag{1.1}$$

where $\boldsymbol{x}$, denoting the activity of devices, is a sparse vector due to the sporadic traffic pattern. Then compressed sensing techniques can be applied to recover the sparse vector. Such grant-free random access has received lots of attention recently. To be specific, when the channel state sequences (recall CES in Fig. 1.1) are contained in the metadata, a joint device activity and data detection problem was studied in [39]. Regarding the sparse linear model proposed in [39], the matrix $\boldsymbol{A}$ in (1.1) captures the channel estimation sequences and pilot sequences, and the vector $\boldsymbol{x}$ in (1.1) represents the information symbols of all devices, where the value is 0 for each inactive device. To improve the efficiency, the overhead caused by metadata has been further reduced. When the CES in Fig. 1.1 is eliminated from the metadata during the packet transmission [33, 36, 37], performed joint channel and data estimation based on various compressed sensing techniques with $\boldsymbol{A}$ in (1.1) capturing the data for all (active and inactive) devices and pilot sequences. Moreover, device activity detection and channel estimation were jointly achieved in the work [38]. In this scenario, the matrix $\boldsymbol{A}$ in (1.1) characterizes the pilot sequences, and the vector $\boldsymbol{x}$ in (1.1) contains the device activity and channel information.

1.1.2 Pilot-Free Communications

In the grant-free random access scheme, the pilot sequence is needed for activity detection, which requires extra bandwidth and induces additional overhead. A more aggressive approach is the pilot-free communication scheme that removes both the fields H_1 and CES in Fig. 1.1 from the metadata. To elude the pilot sequences, more powerful signal processing techniques are needed for data detection. Specifically, a blind demixing based approach has been developed in [11, 14, 24, 25]. Consider an IoT network with one BS and s devices. Each device transmits an encoded signal $\boldsymbol{f} = \boldsymbol{A}\boldsymbol{x}$ to the BS through the channel $\boldsymbol{g}$, where $\boldsymbol{x}$ is the message and $\boldsymbol{A}$ is the encoding matrix, and the received signal at the BS is represented by the cyclic convolution operator $\circledast$,

$$\boldsymbol{y} = \sum_{i=1}^{s} \boldsymbol{f}_i \circledast \boldsymbol{g}_i, \tag{1.2}$$

which is a *blind demixing model* that facilitates to demix the original signals $\{\boldsymbol{f}_i\}$ from the observation $\boldsymbol{y}$ without the knowledge of the channel states $\{\boldsymbol{g}_i\}$. The blind demixing based approach can achieve joint data decoding (i.e., recover data $\boldsymbol{x}$ for each device) and channel estimation. With pilot-free communication supported by blind demixing, the overhead during the transmission is effectively reduced via waiving both the pilot sequences and channel estimation sequences from the metadata.

Considering the sporadic traffic pattern in the IoT network where only part (denoted as $\mathscr{S}$) of the devices are active, a sparse blind demixing model is further developed in [12, 18], given by a *sparse blind demixing model*:

$$y = \sum_{i \in \mathscr{S}} f_i \circledast g_i. \tag{1.3}$$

The estimation for the sparse blind demixing model aims to achieve joint device activity detection and data decoding without the channel state information. Similar to the blind demixing model, it can facilitate to reduce the overhead caused by the channel state information and pilot sequence in the metadata, using more sophisticated detection algorithms.

1.1.3 Identification-Free Communications

Besides the above methods of reducing overhead, excluding the identification information is an important consideration in some IoT applications. Specifically, the identification-free communication scheme eliminates the field H_2 in Fig. 1.1 from the metadata. As an example, suppose that multiple sensors are deployed to take measurements of an unknown parameter vector $\boldsymbol{x}$. In this case, the overhead is mainly dominated by the identity information contained in the metadata [21]. To reduce the overhead, a shuffled linear regression model has been developed. It is established by introducing an unknown permutation matrix $\boldsymbol{\Pi}$ of which the i-th row is the canonical vector $\boldsymbol{e}_{\pi(i)}^{\top}$ of all zeros except a 1 at position $\pi(i)$:

$$\boldsymbol{y} = \boldsymbol{\Pi} \boldsymbol{A} \boldsymbol{x}. \tag{1.4}$$

The goal of the data fusion is to recover $\boldsymbol{x}$ from the permuted data $\boldsymbol{y}$ based on the known sensing matrix $\boldsymbol{A}$. That is, the identities of the signals sent by the sensors are not accessible to the fusion. To address this challenging problem, a line of literatures have developed advanced algorithms from theoretical and practical points of view [30, 31, 34, 35].

1.2 Structured Signal Processing

The techniques mentioned above to achieve low-overhead communications rely on structured signal processing, which exploits underlying structures of the signals or systems, e.g., sparsity, low-rankness, group sparsity or permutation, for effective signal estimation and detection. In this section, we first take a basic structured signal processing problem, i.e., compressed sensing, as an example, to illustrate the main design principles. Then general structured signal processing techniques are introduced.

1.2.1 Example: Compressed Sensing

The key point of compressed sensing is to recover a sparse signal from very few linear measurements. Mathematically, given a *sensing matrix*, i.e., $\boldsymbol{A} \in \mathbb{R}^{m\times n}$, the *compressed sensing problem* can be formulated as recovering $\boldsymbol{x} \in \mathbb{R}^n$ from the observation of

$$\boldsymbol{y} = \boldsymbol{A}\boldsymbol{x} \in \mathbb{R}^m, \tag{1.5}$$

based on the assumption that $\boldsymbol{x}$ has very few nonzero elements, i.e., the ℓ_0-norm $\|\boldsymbol{x}\|_0$ is small. In the sequel, we introduce three key ingredients of a compressed sensing problem: recovery algorithms, measurement mechanisms, and theoretical guarantees.

It is intuitive to recover $\boldsymbol{x}$ from the observation $\boldsymbol{y}$ via solving

$$\underset{\boldsymbol{x}\in\mathbb{C}^n}{\text{minimize}} \quad \|\boldsymbol{x}\|_0 \quad \text{subject to} \quad \boldsymbol{y} = \boldsymbol{A}\boldsymbol{x}. \tag{1.6}$$

The paper [5] showed that problem (1.6) enables to recover a k-sparse signal exactly with a high probability with only $m = k + 1$ random measurements from a Gaussian distributed sensing matrix. Unfortunately, problem (1.6) is a combinatorial optimization problem with an excessive complexity if solved by enumeration [28]. Thus, the tightest convex norm of ℓ_0-norm, i.e., the ℓ_1-norm, is proposed to relax ℓ_0-norm [8], which leads to

$$\underset{\boldsymbol{x}\in\mathbb{C}^n}{\text{minimize}} \quad \|\boldsymbol{x}\|_1 \quad \text{subject to} \quad \boldsymbol{y} = \boldsymbol{A}\boldsymbol{x}. \tag{1.7}$$

Intuitively, this ℓ_1 minimization formulation facilitates to induce sparsity due to the shape of the ℓ_1 ball.

There have been various types of algorithms developed for different formulations of sparse recovery. The most commonly used formulation is the convex relaxation based on (1.7). Given a certain parameter $\lambda > 0$, problem (1.7) can also be represented as an unconstrained optimization problem,

$$\underset{\boldsymbol{x}\in\mathbb{C}^n}{\text{minimize}} \; \frac{1}{2}\|\boldsymbol{A}\boldsymbol{x} - \boldsymbol{y}\|_2^2 + \lambda\|\boldsymbol{x}\|_1. \tag{1.8}$$

Various algorithms have been developed to solve problem (1.8), including interior-point methods [7], projected gradient methods [17], iterative thresholding [10], and the approximate message passing algorithm [15].

Besides effective recovery algorithms, there exist rigorous theoretical guarantees on the recovery of sparse signals, based on specific conditions of the measurements matrix. In particular, the restricted isometry property (RIP) of a sensing matrix $\boldsymbol{A} \in \mathbb{R}^{m\times n}$ was introduced in [7] that measures the degree to which each subset of k

column vectors of $\boldsymbol{A} \in \mathbb{R}^{m \times n}$ is close to being an isometry. A typical theoretical result based on RIP analysis is stated as follows:

Example 1.1 If there exists a $\delta_k \in (0, 1)$ such that the sensing matrix $\boldsymbol{A}$ satisfies

$$(1 - \delta_k) \|\boldsymbol{x}\|_2^2 \leq \|\boldsymbol{Ax}\|_2^2 \leq (1 + \delta_k) \|\boldsymbol{x}\|_2^2 \tag{1.9}$$

for any $\boldsymbol{x}$ that belongs to the set of k-sparse vectors, then problem (1.7) can facilitate to exactly recover the sparse vector $\boldsymbol{x}$ with high probability, provided the number of measurements $m \gtrsim \delta_k^{-2} k \log(n/k)$.

In addition, the exact location of the phase transition for problem (1.7) can be obtained based on conic geometry theory, where a parameter, called the statistical dimension, is introduced to capture the dimension of a linear subspace to the set of convex cones [2]. It demonstrates that under the assumption of i.i.d. standard normal measurements, the transition occurs where the number of measurements, i.e., m, equals the statistical dimension of the descent cone. The shift from failure to success occurs over a range of about $\mathcal{O}(\sqrt{n})$ measurements.

1.2.2 General Structured Signal Processing

Compressive sensing techniques have been successfully applied in many application domains, which have inspired lots of interest in exploiting structures other than sparsity [4, 9, 13, 14, 20, 26, 32, 36]. In the following, we give a brief introduction of the structured signal processing approaches that will be applied for low-overhead communications in this monograph. A general structured signal processing problem with a vector variable is given by

$$\underset{\boldsymbol{x} \in \mathcal{D}_v}{\text{minimize}}\; f(\mathscr{A}\boldsymbol{x}), \tag{1.10}$$

where $\mathscr{A}$ is a linear operator representing the measurement mechanism, f is a loss function, and $\mathcal{D}_v$ is a space of structured vectors (e.g., sparse vectors). For example, in the sparse linear model in (1.1), the operator $\mathscr{A}$ captures the set of pilot matrices and $\boldsymbol{x}$ is a sparse vector. In the shuffled linear regression problem (1.4), the operator $\mathscr{A}$ indicates the permuted sensing matrix. Likewise, a general structured signal processing problem with a matrix variable is given by

$$\underset{\boldsymbol{X} \in \mathcal{D}_m}{\text{minimize}}\; f(\mathscr{A}\boldsymbol{X}), \tag{1.11}$$

where $\mathcal{D}_m$ is a space of structured matrices (e.g., low-rank matrices, sparse matrices, low-rank and sparse matrices, etc.). Specifically, in the blind demixing model (1.2), $\boldsymbol{X}$ is a collection of rank-1 matrices, while $\boldsymbol{X}$ in the sparse blind demixing (1.3) is a collection of low-rank matrices endowed with group sparsity.

The convex program based on the norm operator is typically an effective way to solve problems (1.10) and (1.11) with $\boldsymbol{x}$ and $\boldsymbol{X}$ enjoying certain structures such as sparsity, low-rankness, and group sparsity. Specifically, the convex program with a vector variable can be represented as

$$\underset{\boldsymbol{x}}{\text{minimize}}\, f(\mathscr{A}\boldsymbol{x}), \quad \text{subject to} \quad \|\boldsymbol{x}\|_1 \leq \alpha. \tag{1.12}$$

Here, the ℓ_1-norm can be used for inducing sparsity of a vector. Moreover, the convex program with a matrix variable is given by

$$\underset{\boldsymbol{X}}{\text{minimize}}\, f(\mathscr{A}\boldsymbol{X}), \quad \text{subject to} \quad \mathscr{M}(\boldsymbol{X}) \leq \alpha, \tag{1.13}$$

where the operator $\mathscr{M}(\cdot)$ indicates a norm operator to induce low-rankness, group sparsity, or simultaneous low-rankness and group sparsity. The convex programs (1.12) and (1.13) can be solved via semidefinite programs. Considering the computational complexity and the scalability of the convex program, it motivates to develop nonconvex algorithms that enjoy lower computational complexity. This monograph provides a comprehensive discussion on various algorithms for solving structured signal estimation problems for low-overhead communications from both computational and theoretical points of view.

1.3 Outline

This monograph aims at providing an introduction to key models, algorithms, and theoretical results of structured signal processing in achieving low-overhead communications in IoT networks. Specifically, the content is organized according to four clearly defined categories, i.e., the sparse linear model, blind demixing, sparse blind demixing, and shuffled linear regression, which are summarized in Fig. 1.3. Key problems of low-overhead communications to be considered are also shown in Fig. 1.3. Detailed discussions are provided on methods for solving the above mentioned structured signal processing problems, including convex relaxation approaches, nonconvex approaches, and other optimization algorithms. Moreover, a significant part in each chapter is devoted to statistical theory, demonstrating how to set the corresponding algorithms on solid theoretical foundations, which includes conic integral geometry, algebraic geometric, and Riemannian optimization theory. The proofs of some key results are also included in order to illustrate the theoretical building blocks. For the ease of reference, a brief summary of different models introduced in this monograph and corresponding theory and algorithms is provided in Table 1.1. Moreover, Chap. 6 will introduce the latest developments in learning augmented methods for structured signal processing.

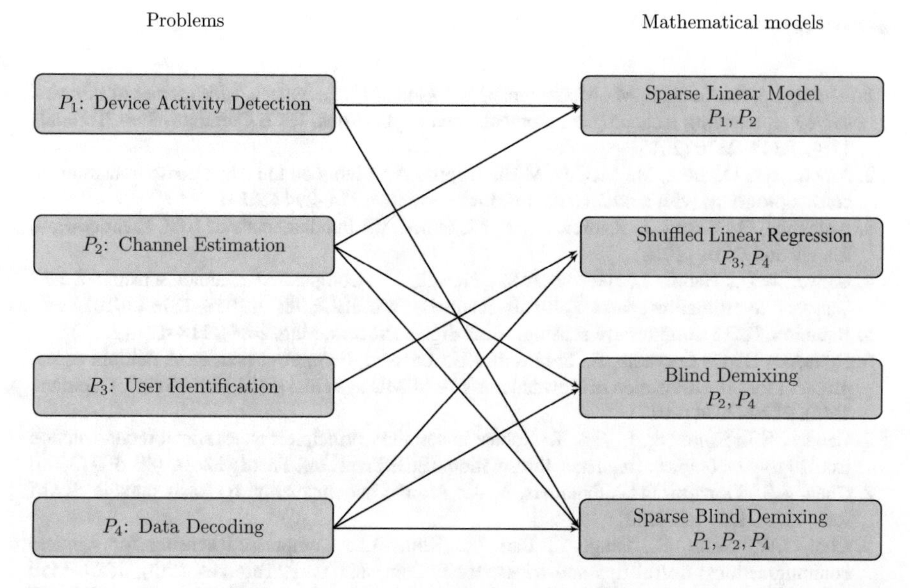

Fig. 1.3 A schematic plot showing the mathematical models and corresponding problems

Table 1.1 Summary of different models, applications, and corresponding theory and algorithms

Model	Application	Formulation	Method (M), theory (T), and algorithm (A)
Sparse linear model	Device activity detection and channel estimation	Model: (2.3) Problem: (2.9)	M: convex relaxation (2.10) T: conic integral geometry
			M: iterative thresholding A: approximate message passing
Blind demixing	Data decoding and channel estimation	Model: (3.13) Problem: (3.20)	M: convex relaxation: (3.20) T: restricted isometry property
			M: nonconvex A: Riemannian trust-region (3.41) Wirtinger flow (regularized (3.22), regularization-free (3.28))
Sparse blind demixing	Device activity detection and data decoding and channel estimation	Model: (4.2) Problem: (4.6)	M: convex relaxation (4.8)
			M: difference-of-convex-functions approach (4.18) A: majorization minimization
			M: smoothed Riemannian optimization (4.30)
Shuffled linear regression	Device activity detection and data decoding	Model: (5.8) Problem: (5.9)	M: maximum likelihood estimation (5.9) A: algorithm based on sorting, approximation algorithm
			M: algebraic–geometric approach (5.31) T: algebraic geometry

References

1. Al-Fuqaha, A., Guizani, M., Mohammadi, M., Aledhari, M., Ayyash, M.: Internet of things: a survey on enabling technologies, protocols, and applications. IEEE Commun. Surv. Tutorials **17**(4), 2347–2376 (2015)
2. Amelunxen, D., Lotz, M., McCoy, M.B., Tropp, J.A.: Living on the edge: phase transitions in convex programs with random data. Inf. Inference **3**(3), 224–294 (2014)
3. Arunabha, G., Zhang, J., Andrews, J.G., Muhamed, R.: Fundamentals of LTE. Prentice-Hall, Englewood Cliffs (2010)
4. Bajwa, W.U., Haupt, J., Sayeed, A.M., Nowak, R.: Compressed channel sensing: a new approach to estimating sparse multipath channels. Proc. IEEE **98**(6), 1058–1076 (2010)
5. Baraniuk, R.G.: Compressive sensing. IEEE Signal Process. Mag. **24**(4), 118–121 (2007)
6. Björnson, E., De Carvalho, E., Sørensen, J.H., Larsson, E.G., Popovski, P.: A random access protocol for pilot allocation in crowded massive MIMO systems. IEEE Trans. Wirel. Commun. **16**(4), 2220–2234 (2017)
7. Candès, E.J., Romberg, J., Tao, T.: Robust uncertainty principles: exact signal reconstruction from highly incomplete frequency information. IEEE Trans. Inf. Theory **52**(2), 489–509 (2006)
8. Chen, S.S., Donoho, D.L., Saunders, M.A.: Atomic decomposition by basis pursuit. SIAM Rev. **43**(1), 129–159 (2001)
9. Choi, J.W., Shim, B., Ding, Y., Rao, B., Kim, D.I.: Compressed sensing for wireless communications: useful tips and tricks. IEEE Commun. Surv. Tutorials **19**(3), 1527–1550 (2017)
10. Daubechies, I., Defrise, M., De Mol, C.: An iterative thresholding algorithm for linear inverse problems with a sparsity constraint. Commun. Pure Appl. Math. **57**(11), 1413–1457 (2004)
11. Dong, J., Shi, Y.: Nonconvex demixing from bilinear measurements. IEEE Trans. Signal Process. **66**(19), 5152–5166 (2018)
12. Dong, J., Shi, Y., Ding, Z.: Sparse blind demixing for low-latency signal recovery in massive IoT connectivity. In: Proceedings of the IEEE International Conference on Acoustics Speech Signal Process (ICASSP), pp. 4764–4768. IEEE, Piscataway (2019)
13. Dong, J., Shi, Y., Ding, Z.: Sparse blind demixing for low-latency signal recovery in massive IoT connectivity. In: Proceedings of the IEEE International Conference on Acoustics Speech Signal Process (ICASSP), pp. 4764–4768 (2019)
14. Dong, J., Yang, K., Shi, Y.: Blind demixing for low-latency communication. IEEE Trans. Wirel. Commun. **18**(2), 897–911 (2019)
15. Donoho, D.L., Maleki, A., Montanari, A.: Message-passing algorithms for compressed sensing. Proc. Natl. Acad. Sci. **106**(45), 18914–18919 (2009)
16. Durisi, G., Koch, T., Popovski, P.: Toward massive, ultrareliable, and low-latency wireless communication with short packets. Proc. IEEE **104**(9), 1711–1726 (2016)
17. Figueiredo, M.A., Nowak, R.D., Wright, S.J.: Gradient projection for sparse reconstruction: application to compressed sensing and other inverse problems. IEEE J. Sel. Top. Sign. Proces. **1**(4), 586–597 (2007)
18. Fu, M., Dong, J., Shi, Y.: Sparse blind demixing for low-latency wireless random access with massive connectivity. In: Proceedings of the IEEE Vehicular Technology Conference (VTC), pp. 4764–4768. IEEE, Piscataway (2019)
19. Hasan, M., Hossain, E., Niyato, D.: Random access for machine-to-machine communication in LTE-advanced networks: issues and approaches. IEEE Commun. Mag. **51**(6), 86–93 (2013)
20. Jiang, T., Shi, Y., Zhang, J., Letaief, K.B.: Joint activity detection and channel estimation for IoT networks: phase transition and computation-estimation tradeoff. IEEE Internet Things J. **6**(4), 6212–6225 (2018)
21. Keller, L., Siavoshani, M.J., Fragouli, C., Argyraki, K., Diggavi, S.: Identity aware sensor networks. In: IEEE INFOCOM, pp. 2177–2185. IEEE, Piscataway (2009)

22. Kong, L., Khan, M.K., Wu, F., Chen, G., Zeng, P.: Millimeter-wave wireless communications for IoT-cloud supported autonomous vehicles: overview, design, and challenges. IEEE Commun. Mag. **55**(1), 62–68 (2017)
23. Letaief, K.B., Chen, W., Shi, Y., Zhang, J., Zhang, Y.A.: The roadmap to 6G: AI empowered wireless networks. IEEE Commun. Mag. **57**(8), 84–90 (2019)
24. Ling, S., Strohmer, T.: Blind deconvolution meets blind demixing: algorithms and performance bounds. IEEE Trans. Inf. Theory **63**(7), 4497–4520 (2017)
25. Ling, S., Strohmer, T.: Regularized gradient descent: a nonconvex recipe for fast joint blind deconvolution and demixing. Inf. Inference J. IMA **8**(1), 1–49 (2019)
26. Liu, L., Larsson, E.G., Yu, W., Popovski, P., Stefanovic, C., De Carvalho, E.: Sparse signal processing for grant-free massive connectivity: a future paradigm for random access protocols in the Internet of Things. IEEE Signal Process. Mag. **35**(5), 88–99 (2018)
27. Motlagh, N.H., Bagaa, M., Taleb, T.: UAV-based IoT platform: a crowd surveillance use case. IEEE Commun. Mag. **55**(2), 128–134 (2017)
28. Muthukrishnan, S., et al.: Data streams: algorithms and applications. Found. Trends Theor. Comput. Sci. **1**(2), 117–236 (2005)
29. Nitsche, T., Cordeiro, C., Flores, A.B., Knightly, E.W., Perahia, E., Widmer, J.: IEEE 802.11 ad: directional 60 GHz communication for multi-Gigabit-per-second Wi-Fi. IEEE Commun. Mag. **52**(12), 132–141 (2014)
30. Pananjady, A., Wainwright, M.J., Courtade, T.A.: Linear regression with shuffled data: statistical and computational limits of permutation recovery. IEEE Trans. Inf. Theory **64**(5), 3286–3300 (2018)
31. Peng, L., Song, X., Tsakiris, M.C., Choi, H., Kneip, L., Shi, Y.: Algebraically-initialized expectation maximization for header-free communication. In: Proceedings of the IEEE International Conference on Acoustics, Speech and Signal Processing (ICASSP), pp. 5182–5186. IEEE, Piscataway (2019)
32. Qin, Z., Fan, J., Liu, Y., Gao, Y., Li, G.Y.: Sparse representation for wireless communications: a compressive sensing approach. IEEE Signal Process. Mag. **35**(3), 40–58 (2018)
33. Schepker, H.F., Bockelmann, C., Dekorsy, A.: Exploiting sparsity in channel and data estimation for sporadic multi-user communication. In: Proceedings of the International Symposium on Wireless Communication Systems, pp. 1–5. VDE, Frankfurt (2013)
34. Tsakiris, M.C., Peng, L.: Homomorphic sensing. In: Proceedings of the International Conference on Machine Learning (ICML), pp. 6335–6344 (2019)
35. Tsakiris, M.C., Peng, L., Conca, A., Kneip, L., Shi, Y., Choi, H.: An algebraic-geometric approach to shuffled linear regression (2018). arXiv:1810.05440
36. Wunder, G., Boche, H., Strohmer, T., Jung, P.: Sparse signal processing concepts for efficient 5G system design. IEEE Access **3**, 195–208 (2015)
37. Wunder, G., Jung, P., Wang, C.: Compressive random access for post-LTE systems. In: Proceedings of the IEEE International Conference on Communications Workshops (ICC), pp. 539–544. IEEE, Piscataway (2014)
38. Xu, X., Rao, X., Lau, V.K.: Active user detection and channel estimation in uplink CRAN systems. In: Proceedings of the IEEE International Conference on Communications (ICC), pp. 2727–2732. IEEE, Piscataway (2015)
39. Zhu, H., Giannakis, G.B.: Exploiting sparse user activity in multiuser detection. IEEE Trans. Commun. **59**(2), 454–465 (2010)

Chapter 2
Sparse Linear Model

Abstract In this chapter, a sparse linear model for joint activity detection and channel estimation in IoT networks is introduced. We present the problem formulation for both the cases of single-antenna and multiple-antenna BSs. A convex relaxation approach based on ℓ_p-norm minimization is firstly introduced, followed by a smoothed primal-dual first-order algorithm to solve it. The theoretical analysis of the convex relaxation approach based on the conic integral geometry theory is further presented. Furthermore, an iterative threshold algorithm, namely approximate message passing (AMP), is introduced, followed by the performance analysis based on the state evolution technique. Simulation results are also presented to demonstrate the performance of different algorithms.

2.1 Joint Activity Detection and Channel Estimation

Under the grant-free random access scheme, the metadata contains control information, e.g., the user identifier and pilot for channel estimation, and payload data that are transmitted together to the BS [24, 25]. Due to the finite channel coherence time and the massive number of devices in the IoT network, it is impossible to assign orthogonal pilot sequences to different devices [26]. Moreover, incorporating a separate pilot sequence for channel estimation in the metadata would bring redundant overheads. Considering the typical small payload size of IoT applications, it is of vital importance to reduce the overheads.

One unique characteristic of massive IoT connectivity is the sporadic data traffic, i.e., only a part of devices in the network are active at each time slot. This is because IoT devices are often designed to be in the sleep mode most of the time to conserve energy and are only triggered by external events to transmit data [26]. Exploiting this fact, a *sparse linear model* can well capture the problem of massive connectivity, which enables joint device activity detection and channel estimation [26]. This structured model describes an underdetermined linear system with more

Y. Shi et al., *Low-overhead Communications in IoT Networks*,
https://doi.org/10.1007/978-981-15-3870-4_2

unknown variables than equations. Considering an IoT network consisting of one BS and N devices, a sparse linear model can be established as

$$\boldsymbol{y} = \boldsymbol{A}\boldsymbol{x}, \tag{2.1}$$

where $\boldsymbol{y} \in \mathbb{R}^L$ is the received signal at the BS, $\boldsymbol{A} \in \mathbb{R}^{L\times N}$ is the set of pilot sequences, and $\boldsymbol{x} \in \mathbb{R}^N$ is a sparse vector containing the information of the activity states of devices and the channel states. Particularly, $\boldsymbol{A}$ is chosen from a set of non-orthogonal preamble sequences. The corresponding element of $\boldsymbol{x}$ is 0 for an inactive device; otherwise, it denotes the channel coefficient for an active device. Therefore, by recovering the sparse vector $\boldsymbol{x}$ from the observation $\boldsymbol{y}$, device activity detection and channel estimation can be simultaneously achieved.

To solve the estimation problem (2.1), the work [36] proposed a modified Bayesian compressed sensing algorithm. To further improve the performance of the algorithm, the works [31, 34, 35] developed the AMP algorithm with the performance analysis based on the fading coefficients and statistical channel information. The rigorous analysis has been recently investigated in a line of literatures. It shows that a state evolution analysis [3, 24, 25] of the AMP algorithm enables to characterize the false alarm and miss detection probabilities for activity detection. Recently, the paper [21] has developed a structured group sparsity estimation approach to achieve joint device activity detection and channel estimation. To increase the convergence rate and guarantee the accuracy, a smoothing method has been proposed in [21] to solve the group sparsity estimation problem, and sharp computation and estimation trade-offs of this method were further provided [21].

In the following, we first illustrate how the sparse linear model helps to formulate the joint activity detection and channel estimation problem. Then, effective algorithms and rigorous analysis are provided, and both convex and nonconvex approaches are considered.

2.2 Problem Formulation

This section presents the problem formulation for joint activity detection and channel estimation, for both single-antenna and multi-antenna BSs. Assume there is one BS along with N devices in an IoT network. Due to sporadic traffic, only a part of the devices are active in each time slot. For each coherent block in a synchronized wireless system with block fading, the indicator function that implies the device activity is defined as:

$$\alpha_i = \begin{cases} 1, & \text{if device } i \text{ is active,} \\ 0, & \text{otherwise,} \end{cases} \quad \forall i \in \{1, \ldots, N\}. \tag{2.2}$$

Hence, $\mathcal{S} = \{i \mid \alpha_i = 1, i = 1, \ldots, N\}$ denotes the set of active devices within a coherence block, with the number of active devices being $|\mathcal{S}|$.

2.2.1 Single-Antenna Scenario

Assume that the BS is equipped with a single antenna, and denote the channel coefficient from device i to the BS as h_i for $i = 1, \ldots, N$. Define $\boldsymbol{q}_i \in \mathbb{C}^L$ as the pilot sequence transmitted from device i, where L is the length of the pilot sequence which is much smaller than the number of devices, i.e., $L \ll N$, due to the finite coherence time. The received signal over L symbols at the BS is given by

$$\boldsymbol{y} = \sum_{i=1}^{N} \alpha_i h_i \boldsymbol{q}_i + \boldsymbol{n} = \sum_{i \in \mathcal{S}} h_i \boldsymbol{q}_i + \boldsymbol{n} = \boldsymbol{A}\boldsymbol{x} + \boldsymbol{n}, \tag{2.3}$$

where $\boldsymbol{y} = [y_1, \ldots, y_L]^\top \in \mathbb{C}^L$ is the received signal, $q_{i,\ell} \sim \mathcal{CN}(0, 1) \in \mathbb{C}$ for $i = 1, \ldots, N$, $\ell = 1, \ldots, L$ are pilot symbols, and $\boldsymbol{n} \in \mathbb{C}^L \sim \mathcal{CN}(\boldsymbol{0}, \sigma^2 \boldsymbol{I})$ is the additive white Gaussian noise. Moreover,

$$\boldsymbol{A} = [\boldsymbol{q}_1, \ldots, \boldsymbol{q}_N] \in \mathbb{C}^{L \times N}$$

is the collection of pilot sequences of all the devices, and

$$\boldsymbol{x} = [x_1, \ldots, x_N]^\top \in \mathbb{C}^N$$

with $x_i = \alpha_i h_i$ for $i = 1, \ldots, N$ contain device activity indicators and channel states. Here, Eq. (2.3) gives a sparse linear model. The task for the BS is to jointly detect the active devices and estimate the channel coefficients by recovering $\boldsymbol{x}$ from the observation $\boldsymbol{y}$, which can then be used for data detection. An example of the sparse linear model is illustrated in Example 2.1.

Example 2.1 Consider a network with two devices and one BS equipped with a single antenna. Assume that the pilot sequences $\boldsymbol{A}$ are predefined as:

$$\boldsymbol{A} = \begin{bmatrix} 1 & -\sqrt{3} \\ 1 & \sqrt{3} \\ 2 & 0 \end{bmatrix}. \tag{2.4}$$

Assuming that the second device is inactive and the channel state of the active device (i.e., device 1) is $h_1 = 1$, we have

$$\boldsymbol{x} = \begin{bmatrix} 1 \cdot h_1 \\ 0 \cdot h_2 \end{bmatrix} = \begin{bmatrix} 1 \\ 0 \end{bmatrix}. \tag{2.5}$$

It yields a sparse linear model:

$$\boldsymbol{y} = \boldsymbol{A}\boldsymbol{x} = \begin{bmatrix} 1 & -\sqrt{3} \\ 1 & \sqrt{3} \\ 2 & 0 \end{bmatrix} \begin{bmatrix} 1 \\ 0 \end{bmatrix} = \begin{bmatrix} 1 \\ 1 \\ 2 \end{bmatrix}. \tag{2.6}$$

2.2.2 Multiple-Antenna Scenario

Inspired by the successful application of the sparse linear model for device activity detection, multi-antenna technologies have been applied to enhance the detection performance. It generalizes the sparse signal-recovery problem to the case with a group of measurement vectors. These signal vectors are assumed to be sparse and share a common support, corresponding to the active devices. This induces a group sparsity structure, which helps to improve the performance of device activity detection and channel estimation.

Assume the BS is equipped with M antennas. The ℓ-th received signal at the BS is denoted as $\boldsymbol{y}(\ell) \in \mathbb{C}^M$ for all $\ell = 1, \ldots, L$, which is given by

$$\boldsymbol{y}(\ell) = \sum_{i=1}^{N} \boldsymbol{h}_i \alpha_i q_i(\ell) + \boldsymbol{n}(\ell) = \sum_{i \in \mathscr{S}} \boldsymbol{h}_i q_i(\ell) + \boldsymbol{n}(\ell), \tag{2.7}$$

for all $\ell = 1, \ldots, L$. Here, $q_i(\ell) \sim \mathscr{CN}(0, 1) \in \mathbb{C}$ is the pilot symbol transmitted from device i at time slot ℓ, $\boldsymbol{h}_i \in \mathbb{C}^M$ denotes the channel vector from device i to the BS antennas, and $\boldsymbol{n}(\ell) \in \mathbb{C}^M \sim \mathscr{CN}(\boldsymbol{0}, \sigma^2 \boldsymbol{I})$ is the independent additive white Gaussian noise.

By accumulating the signal vectors over L time slots, we get the aggregated received signal matrix

$$\boldsymbol{Y} = [\boldsymbol{y}(1), \ldots, \boldsymbol{y}(L)]^\top \in \mathbb{C}^{L \times M},$$

the channel matrix

$$\boldsymbol{H} = [\boldsymbol{h}_1, \ldots, \boldsymbol{h}_N]^\top \in \mathbb{C}^{N \times M},$$

the additive noise matrix

$$\boldsymbol{N} = [\boldsymbol{n}(1), \ldots, \boldsymbol{n}(L)] \in \mathbb{C}^{L \times M},$$

and pilot matrix

$$\boldsymbol{Q} = [\boldsymbol{q}(1), \ldots, \boldsymbol{q}(L)]^\top \in \mathbb{C}^{L \times N},$$

where $\boldsymbol{q}(\ell) = [q_1(\ell), \ldots, q_N(\ell)]^\top \in \mathbb{C}^N$. Thus, (2.7) can be rewritten as

$$\boldsymbol{Y} = \boldsymbol{Q}\boldsymbol{\Theta} + \boldsymbol{N}, \tag{2.8}$$

where the matrix $\boldsymbol{\Theta}$ is given by $\boldsymbol{\Theta} = \boldsymbol{D}\boldsymbol{H} \in \mathbb{C}^{N \times M}$ with $\boldsymbol{D} = \mathrm{diag}(\alpha_1, \ldots, \alpha_n) \in \mathbb{R}^{N \times N}$ being the diagonal activity matrix. Hence, the matrix $\boldsymbol{\Theta}$ endows with a group sparse structure. The task for the multi-antenna BS is to detect the active devices and estimate the channel matrix by recovering $\boldsymbol{\Theta}$ from the observation $\boldsymbol{Y}$.

2.3 Convex Relaxation Approach

2.3.1 Method: ℓ_p-Norm Minimization

In this section, we present a convex relaxation approach for joint device activity detection and channel estimation. For the noiseless case, the straightforward idea of recovering a sparse signal $\boldsymbol{x}$ of which most elements are zeros is to find the sparsest signal among all those that generate the observation $\boldsymbol{y} = \boldsymbol{A}\boldsymbol{x}$. It results in the following problem:

$$\begin{aligned} &\underset{\boldsymbol{x}\in\mathbb{C}^N}{\text{minimize}} \quad \|\boldsymbol{x}\|_0 \\ &\text{subject to} \quad \boldsymbol{y} = \boldsymbol{A}\boldsymbol{x}, \end{aligned} \tag{2.9}$$

where the ℓ_0-norm describes the number of nonzeros in $\boldsymbol{x}$. However, the problem is NP-hard due to the inevitable combinatorial search [27]. A convex relaxation approach can be applied by replacing the ℓ_0-norm by the ℓ_1-norm. The method of ℓ_1-norm minimization [8, 10, 14], which exploits ℓ_1-norm to induce the sparsity of the signal $\boldsymbol{x}$, is a well-established approach to solve compressed sensing problems.

The optimization problem that recovers $\boldsymbol{x}$ from the noisy observation $\boldsymbol{y}$ in (2.3) is formulated as

$$\begin{aligned} &\underset{\boldsymbol{x}\in\mathbb{C}^N}{\text{minimize}} \quad \|\boldsymbol{x}\|_1 \\ &\text{subject to} \quad \|\boldsymbol{A}\boldsymbol{x} - \boldsymbol{y}\|_2 \leq \epsilon, \end{aligned} \tag{2.10}$$

where the parameter $\epsilon > 0$ is a prior threshold such that $\boldsymbol{n}$ in (2.3) obeys $\|\boldsymbol{n}\|_2 \leq \epsilon$. Given the estimate vector $\hat{\boldsymbol{x}}$, the activity matrix can be recovered as

$$\hat{\boldsymbol{C}} = \text{diag}(\hat{a}_1, \ldots, \hat{a}_n),$$

where $\hat{a}_i = 1$ if $|\hat{x}_i| \geq \gamma_0$ for a small enough threshold γ_0 ($\gamma_0 \geq 0$); otherwise, $\hat{a}_i = 0$. The estimated channel vector for the active devices is thus given by $\hat{\boldsymbol{h}}$ with its i-th element as $\hat{h}_i = \hat{x}_i$, where $i \in \{j | \hat{a}_j = 1\}$.

Likewise, the optimization problem in the multiple-antenna scenario can be presented as

$$\begin{aligned} &\underset{\boldsymbol{\Theta}\in\mathbb{C}^{N\times M}}{\text{minimize}} \ \mathscr{R}(\boldsymbol{\Theta}) := \sum_{i=1}^{N} \|\boldsymbol{\theta}^i\|_2 \\ &\text{subject to} \ \|\boldsymbol{Q}\boldsymbol{\Theta} - \boldsymbol{Y}\|_F \leq \epsilon, \end{aligned} \tag{2.11}$$

where $\epsilon > 0$ is a priori such that $\boldsymbol{N}$ in (2.8) obeys $\|\boldsymbol{N}\|_F \leq \epsilon$, and $\boldsymbol{\theta}^i$ is the i-th row of matrix $\boldsymbol{\Theta}$. Here the function $\mathscr{R}(\boldsymbol{\Theta})$ induces the group sparsity via mixed ℓ_1/ℓ_2-

norm, where ℓ_2-norm $\|\boldsymbol{\theta}^i\|_2$ bounds the magnitude of the elements of $\boldsymbol{\theta}^i$, while ℓ_1-norm induces the sparsity of $[\|\boldsymbol{\theta}^1\|_2, \ldots, \|\boldsymbol{\theta}^N\|_2]$. Given the estimated matrix $\hat{\boldsymbol{\Theta}}$, the activity matrix can be recovered as $\hat{\boldsymbol{C}} = \text{diag}(\hat{a}_1, \ldots, \hat{a}_n)$, where $\hat{a}_i = 1$ if $\|\hat{\boldsymbol{\theta}}^i\|_2 \geq \gamma_0$ for a small enough threshold $\gamma_0 (\gamma_0 \geq 0)$; otherwise, $\hat{a}_i = 0$. The estimated channel matrix for the active devices is thus given by $\hat{\boldsymbol{H}}$ with its i-th row as $\hat{\boldsymbol{h}}^i = \hat{\boldsymbol{\theta}}^i$, where $i \in \{j | \hat{a}_j = 1\}$.

The convex relaxation approaches can be applied to solve problems (2.10) and (2.11) in polynomial time. However, the general interior point solvers that are typically used to deal with SDP are impractical to be applied in large-scale problems, due to the high computational complexity. It motivates to develop fast, first-order algorithms with reduced computational complexity.

2.3.2 *Algorithm: Smoothed Primal-Dual First-Order Methods*

The first-order methods, e.g., gradient methods, proximal methods [30], alternating direction method of multipliers (ADMM) algorithm [6, 33], fast ADMM algorithm [19], and Nesterov-type algorithms [4], can efficiently solve large-scale problems. Furthermore, one way to lower the computational complexity is to accelerate the convergence rate without increasing the computational cost of each iteration. It was shown in [29] that with a large data size it is possible to increase the step size in the projected gradient method, thereby achieving a faster convergence rate. The paper [18] showed that via adjusting the original iterations, it is possible to achieve faster convergence rates and maintain the estimation accuracy without greatly increasing the computational cost of each iteration. Furthermore, the acceleration of convergence rates can be achieved via smoothing techniques such as convex relaxation [9], or simply adding a smooth function to smooth the non-differentiable objective function [4, 7, 22]. However, the quantity of smoothing should be chosen thoughtfully to guarantee the performance of sporadic device activity detection in IoT networks. To address the limitations above, the paper [21] proposed a smoothed primal-dual first-order method to solve the high-dimensional group sparsity estimation problem. The sharp trade-offs between the computational cost and estimation accuracy are rigorously characterized in [21], which is further discussed in Sect. 2.3.3.2. The smoothing algorithm is first presented in the following.

By adding a smoothing function $\frac{\mu}{2}\|\boldsymbol{\Theta}\|_F^2$, where μ is a positive scalar and called as the smoothing parameter, problem (2.11) is reformulated as

$$\begin{aligned} \underset{\boldsymbol{\Theta} \in \mathbb{C}^{N \times M}}{\text{minimize}} \quad & \tilde{\mathscr{R}}(\boldsymbol{\Theta}) := \mathscr{R}(\boldsymbol{\Theta}) + \frac{\mu}{2}\|\boldsymbol{\Theta}\|_F^2 \\ \text{subject to} \quad & \|\boldsymbol{Q}\boldsymbol{\Theta} - \boldsymbol{Y}\|_F \leq \epsilon. \end{aligned} \tag{2.12}$$

To facilitate algorithm design, the sparse linear observation is represented in the real domain as follows:

$$\begin{aligned}\tilde{Y} &= \tilde{Q}\tilde{\Theta}_0 + \tilde{N}\\ &= \begin{bmatrix}\Re\{Q\} & -\Im\{Q\}\\ \Im\{Q\} & \Re\{Q\}\end{bmatrix}\begin{bmatrix}\Re\{\Theta_0\}\\ \Im\{\Theta_0\}\end{bmatrix} + \begin{bmatrix}\Re\{N\}\\ \Im\{N\}\end{bmatrix}.\end{aligned} \tag{2.13}$$

The function $\tilde{\mathscr{R}}(\boldsymbol{\Theta})$ with respect to the complex matrix $\boldsymbol{\Theta} \in \mathbb{C}^{N\times M}$ can be further converted to the function $\tilde{\mathscr{R}}_G(\tilde{\boldsymbol{\Theta}})$ with respect to the real matrix $\tilde{\boldsymbol{\Theta}} \in \mathbb{R}^{2N\times M}$ as

$$\tilde{\mathscr{R}}_G(\tilde{\boldsymbol{\Theta}}) = \sum_{i=1}^{N} \|\tilde{\boldsymbol{\Theta}}_{\mathscr{V}_i}\|_F + \frac{\mu}{2}\|\tilde{\boldsymbol{\Theta}}_{\mathscr{V}_i}\|_F^2. \tag{2.14}$$

Here

$$\tilde{\boldsymbol{\Theta}}_{\mathscr{V}_i} = [(\tilde{\boldsymbol{\theta}}^i)^\top, (\tilde{\boldsymbol{\theta}}^{i+N})^\top]^\top$$

is the row submatrix of $\tilde{\boldsymbol{\Theta}}$ consisting of the rows indexed by $\mathscr{V}_i = \{i, i+N\}$. Hence, problem (2.12) can be approximated as the following structured group sparse estimation problem

$$\begin{aligned}&\underset{\tilde{\boldsymbol{\Theta}}\in\mathbb{R}^{2N\times M}}{\text{minimize}}\ \tilde{\mathscr{R}}_G(\tilde{\boldsymbol{\Theta}})\\ &\text{subject to}\ \|\bar{\boldsymbol{Q}}\tilde{\boldsymbol{\Theta}} - \tilde{\boldsymbol{Y}}\|_F \le \epsilon,\end{aligned} \tag{2.15}$$

where $\bar{\boldsymbol{Q}} \in \mathbb{R}^{2L\times 2N} \sim \mathcal{N}(\mathbf{0}, 0.5\boldsymbol{I})$ is designed as a Gaussian random matrix. Due to the indifferentiability of problem (2.15), it yields a slow coverage rate when solved by the subgradient method. Fortunately, the dual formulation of problem (2.15) leverages the benefits from smoothing techniques. In particular, the smoothed dual problem can be transferred to an unconstrained problem with the composite objective function consisting of a convex, nonsmooth function and a convex, smooth function. The dual problem of (2.15) is represented as

$$\begin{aligned}&\underset{\boldsymbol{Z},t}{\text{maximize}}\quad \mathscr{D}(\boldsymbol{Z}, t) := \inf_{\tilde{\boldsymbol{\Theta}}}\left\{\tilde{\mathscr{R}}(\tilde{\boldsymbol{\Theta}}) - \langle \boldsymbol{Z}, \tilde{\boldsymbol{Q}}\tilde{\boldsymbol{\Theta}} - \tilde{\boldsymbol{Y}}\rangle - t\epsilon\right\}\\ &\text{subject to}\quad \|\boldsymbol{Z}\|_F \le t,\end{aligned}$$

where $\boldsymbol{Z} \in \mathbb{R}^{2N\times M}$ and $t > 0$. Since the parameter $\epsilon \ge 0$ eludes the dual variable t, it yields the unconstrained problem:

$$\underset{\boldsymbol{Z}\in\mathbb{R}^{2N\times M}}{\text{minimize}}\ \mathscr{D}(\boldsymbol{Z}) := -\inf_{\tilde{\boldsymbol{\Theta}}}\left\{\tilde{\mathscr{R}}(\tilde{\boldsymbol{\Theta}}) - \langle \boldsymbol{Z}, \tilde{\boldsymbol{Q}}\tilde{\boldsymbol{\Theta}} - \tilde{\boldsymbol{Y}}\rangle - \epsilon\|\boldsymbol{Z}\|_F\right\}. \tag{2.16}$$

The dual objective function $\tilde{\mathscr{D}}(\mathbf{Z})$ can be further represented as a composite function

$$\mathscr{D}(\mathbf{Z}) = \tilde{\mathscr{D}}(\mathbf{Z}) + \mathscr{H}(\mathbf{Z}), \tag{2.17}$$

where

$$\tilde{\mathscr{D}}(\mathbf{Z}) = -\inf_{\tilde{\boldsymbol{\Theta}}} \left\{ \tilde{\mathscr{R}}(\tilde{\boldsymbol{\Theta}}) - \langle \mathbf{Z}, \tilde{\boldsymbol{Q}}\tilde{\boldsymbol{\Theta}} \rangle \right\} - \langle \mathbf{Z}, \tilde{\boldsymbol{Y}} \rangle$$

and $\mathscr{H}(\mathbf{Z}) = \epsilon \|\mathbf{Z}\|_F$. The gradient of the function $\tilde{\mathscr{D}}(\mathbf{Z})$ is

$$\nabla \tilde{\mathscr{D}}(\mathbf{Z}) = -\tilde{\boldsymbol{Y}} + \tilde{\boldsymbol{Q}}\tilde{\boldsymbol{\Theta}}_{\mathbf{Z}},$$

where

$$\tilde{\boldsymbol{\Theta}}_{\mathbf{Z}} := \arg\min_{\tilde{\boldsymbol{\Theta}}} \left\{ \tilde{\mathscr{R}}(\tilde{\boldsymbol{\Theta}}) - \langle \mathbf{Z}, \tilde{\boldsymbol{Q}}\tilde{\boldsymbol{\Theta}} \rangle \right\}. \tag{2.18}$$

In addition, $\nabla \tilde{\mathscr{D}}(\mathbf{Z})$ is Lipschitz continuous with the Lipschitz constant being bounded by $L_s := \mu^{-1} \|\tilde{\boldsymbol{Q}}\|_2^2$. The composite form in (2.17) can be solved by a set of first-order approaches [4]. These methods are exceptionally sensitive to the smoothing parameter μ, which means that a larger value of the smoothing parameter μ induces a faster convergence rate. For instance, the Lan, Lu, and Monteiro's algorithm [23] is illustrated in Algorithm 2.1 as a typical example to show the benefits of smoothing.

Algorithm 2.1: Lan, Lu, and Monteiro's algorithm

Input : Pilot matrix $\tilde{\boldsymbol{Q}} \in \mathbb{R}^{2L\times 2N}$, Lipschitz constant $L_s := \mu^{-1}\|\tilde{\boldsymbol{Q}}\|_2^2$, observation matrix $\tilde{\boldsymbol{Y}} \in \mathbb{R}^{2L\times M}$, and parameter ϵ.

1 $\mathbf{Z}_0 \leftarrow \mathbf{0}$, $\bar{\mathbf{Z}}_0 \leftarrow \mathbf{Z}_0$, $t_0 \leftarrow 1$

2 for $k = 0, 1, 2, \ldots$ do

3 $\quad \boldsymbol{B}_k \leftarrow (1 - t_k)\mathbf{Z}_k + t_k \bar{\mathbf{Z}}_k$

4 $\quad \tilde{\boldsymbol{\Theta}}_k \leftarrow \mu^{-1}\text{SoftThreshold}(\tilde{\boldsymbol{Q}}^{\mathrm{T}} \boldsymbol{B}_k, 1)$

5 $\quad \bar{\mathbf{Z}}_{k+1} \leftarrow \text{Shrink}(\bar{\mathbf{Z}}_k - (\tilde{\boldsymbol{Q}}\tilde{\boldsymbol{\Theta}}_k - \tilde{\boldsymbol{Y}})/L_s/t_k, \epsilon/L_s/t_k)$

6 $\quad \mathbf{Z}_{k+1} \leftarrow \text{Shrink}(\boldsymbol{B}_k - (\tilde{\boldsymbol{Q}}\tilde{\boldsymbol{\Theta}}_k - \tilde{\boldsymbol{Y}})/L_s, \epsilon/t_k)$

7 $\quad t_{k+1} \leftarrow 2/(1 + (1 + 4/t_k^2)^{1/2})$

8 end

In Algorithm 2.1, Line 4 is the solution to (2.18), Lines 5 and 6 are the solutions to the following gradient mapping, respectively,

$$\bar{\boldsymbol{Z}}_{k+1} \leftarrow \arg\min_{\boldsymbol{Z}\in\mathbb{R}^{2N\times M}} \left\{ \langle \nabla\tilde{\mathscr{D}}(\boldsymbol{Z}), \boldsymbol{Z}\rangle + \frac{1}{2} t_k L_s \|\boldsymbol{Z} - \bar{\boldsymbol{Z}}_k\|_F + \mathscr{H}(\boldsymbol{Z}) \right\},$$

$$\boldsymbol{Z}_{k+1} \leftarrow \arg\min_{\boldsymbol{Z}\in\mathbb{R}^{2N\times M}} \left\{ \langle \nabla\tilde{\mathscr{D}}(\boldsymbol{Z}), \boldsymbol{Z}\rangle + \frac{1}{2} L_s \|\boldsymbol{Z} - \boldsymbol{B}_k\|_F + \mathscr{H}(\boldsymbol{Z}) \right\}.$$

Denote $\boldsymbol{Z}^*$ as an optimal solution for (2.16), then the convergence behavior of Algorithm 2.1 is demonstrated as [4]

$$\mathscr{D}(\boldsymbol{Z}_{k+1}) - \mathscr{D}(\boldsymbol{Z}^*) \leq \frac{2\|\tilde{\boldsymbol{Q}}\|_2^2 \|\boldsymbol{Z}_0 - \boldsymbol{Z}^*\|_F^2}{\mu k^2}. \tag{2.19}$$

Based on (2.19), the number of iterations

$$\left\lceil \sqrt{2\|\tilde{\boldsymbol{Q}}\|_2^2/(\mu\epsilon_0)} \|\boldsymbol{Z}_0 - \boldsymbol{Z}^*\|_F \right\rceil$$

is required to reach the accuracy of ϵ_0. That is, a larger μ would lead to a faster convergence rate.

2.3.3 *Analysis: Conic Integral Geometry*

The paper [21] discussed the trade-off between the estimation accuracy and computational cost in terms of the smoothing method described in Sect. 2.3.2, which is achieved by characterizing the convergence rate in terms of the smoothing parameter, pilot sequence length, and estimation accuracy. The analysis is based on the theory of conic integral geometry [1, 28, 32]. Prior to focusing on conic integral geometry for the sparse linear model, you may refer to Sect. 8.1 to have a basic overview of conic integral geometry.

2.3.3.1 Conic Integral Geometry for the Sparse Linear Model

Considering the smoothing method illustrated in Sect. 2.3.2, it is critical to find a proper smoothing parameter μ, which can be achieved by analyzing the trade-off between the estimation accuracy and computational cost of the convex optimization problem (2.11). *Conic integral geometry* theory turns out to be a promising and powerful tool to predict phase transitions (including the location and width of the transition region) for random cone programs in the real field case [1, 28, 32]. Based on the conic integral geometry, the paper [21] proposed to approximate the original complex estimation problem (2.11) by a real estimation problem, followed by analyzing on the performance of the proposed smoothing method concerning the smoothing parameter μ.

In the noiseless scenario, we consider the following approximated problem:

$$\begin{aligned} &\underset{\tilde{\boldsymbol{\Theta}} \in \mathbb{R}^{2N \times M}}{\text{minimize}}\ \mathscr{R}_G(\tilde{\boldsymbol{\Theta}}) \\ &\text{subject to}\ \tilde{\boldsymbol{Y}} = \bar{\boldsymbol{Q}}\tilde{\boldsymbol{\Theta}}, \end{aligned} \tag{2.20}$$

where

$$\mathscr{R}_G(\tilde{\boldsymbol{\Theta}}) = \sum_{i=1}^{N} \|\tilde{\boldsymbol{\Theta}}_{\mathcal{V}_i}\|_F$$

and $\tilde{\boldsymbol{\Theta}}$, $\bar{\boldsymbol{Q}}$, and $\tilde{\boldsymbol{Y}}$ are defined in (2.15). To deal with problem (2.20), several definitions and facts in convex analysis [1] are introduced first.

Definition 2.1 (Descent Cone) The descent cone $\mathscr{D}(\mathscr{R}, \boldsymbol{x})$ of a proper convex function $\mathscr{R} : \mathbb{R}^d \rightarrow \mathbb{R} \cup \{\pm\infty\}$ at point $\boldsymbol{x} \in \mathbb{R}^d$ is the conic hull of the perturbations that do not increase $\mathscr{R}$ near $\boldsymbol{x}$, i.e.,

$$\mathscr{D}(\mathscr{R}, \boldsymbol{x}) = \bigcup_{\tau > 0} \left\{ \boldsymbol{y} \in \mathbb{R}^d : \mathscr{R}(\boldsymbol{x} + \tau \boldsymbol{y}) \leq \mathscr{R}(\boldsymbol{x}) \right\}.$$

Fact 2.1 (Optimality Condition) *Let $\mathscr{R}$ be a proper convex function. Matrix $\tilde{\boldsymbol{\Theta}}_0$ is the unique optimal solution to problem (2.20) if and only if*

$$\mathscr{D}(\mathscr{R}_G, \tilde{\boldsymbol{\Theta}}_0) \bigcap \operatorname{null}(\bar{\boldsymbol{Q}}, M) = \{\boldsymbol{0}\},$$

where

$$\operatorname{null}(\bar{\boldsymbol{Q}}, M) = \{\boldsymbol{Z} \in \mathbb{R}^{2N \times M} : \bar{\boldsymbol{Q}}\boldsymbol{Z} = \boldsymbol{0}_{2L \times M}\}$$

denotes the null space of the operator $\bar{\boldsymbol{Q}} \in \mathbb{R}^{2L \times 2N}$.

Figure 2.1 illustrates the geometry of the optimality condition described in Fact 2.1. Specifically, problem (2.20) succeeds to yield optimal solution if and only if the null space of $\bar{\boldsymbol{Q}}$ misses the cone of descent directions of $\mathscr{R}_G$ at the ground truth $\tilde{\boldsymbol{\Theta}}_0$, which is illustrated in Fig. 2.1a; otherwise, it fails to obtain the optimal solution follows $\tilde{\boldsymbol{\Theta}}^* \neq \tilde{\boldsymbol{\Theta}}_0$, which is illustrated in Fig. 2.1b.

To characterize the phase transition in two intersection cones, the concept of statistical dimension is proposed in [1] that is the generalization of the dimension of linear subspaces.

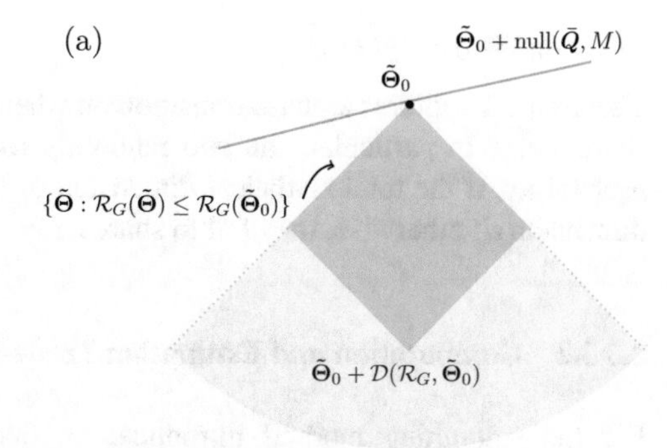

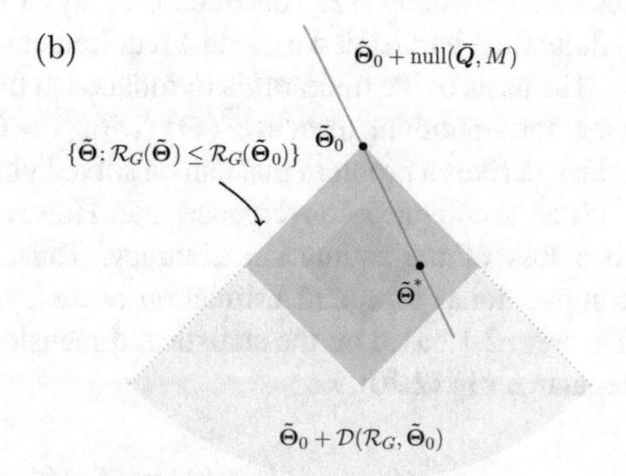

Fig. 2.1 Optimality condition for problem (2.20). (**a**) Problem succeeds. (**b**) Problem fails

Definition 2.2 (Statistical Dimension) The statistical dimension $\delta(C)$ of a closed convex cone C in $\mathbb{R}^d$ is defined as:

$$\delta(C) = \mathbb{E}[\|\boldsymbol{\Pi}_C(\boldsymbol{g})\|_2^2], \tag{2.21}$$

where $\boldsymbol{g} \in \mathbb{R}^d$ is a standard normal vector, and

$$\boldsymbol{\Pi}_C(\boldsymbol{x}) = \arg\min\{\|\boldsymbol{x} - \boldsymbol{y}\|_2 : \boldsymbol{y} \in C\}$$

denotes the Euclidean projection onto C.

The statistical dimension enables to measure the size of convex cones. Based on the statistical dimensions of general convex cones, the approximated conic kinematic formula can be presented as follows [2].

Theorem 2.1 (Approximate Kinematic Formula) *Fix a tolerance* $\eta \in (0, 1)$. *Let* C *and* K *be convex cones in* $\mathbb{R}^d$, *but one of them is not a subspace. Draw a random orthogonal basis* $\boldsymbol{U}$. *Then*

$$\delta(C) + \delta(K) \le d - a_\eta\sqrt{d} \Longrightarrow \mathbb{P}\{C \cap \boldsymbol{U}K \ne \{\boldsymbol{0}\}\} \le \eta$$

$$\delta(C) + \delta(K) \ge d + a_\eta\sqrt{d} \Longrightarrow \mathbb{P}\{C \cap \boldsymbol{U}K \ne \{\boldsymbol{0}\}\} \ge 1 - \eta,$$

where $a_\eta := \sqrt{8\log(4/\eta)}$.

Theorem 2.1 captures a phase transition on whether the two randomly rotated cones share a ray. In particular, the two randomly rotated cones share a ray with high probability, if the total statistical dimension of the two cones exceeds the ambient dimension d; otherwise, they fail to share a ray.

2.3.3.2 Computation and Estimation Trade-Offs

For the smoothing method introduced in Sect. 2.3.2, a trade-off between the computational cost and estimation accuracy is characterized based on the general results in Theorem 2.1. This trade-off plays a vital role in massive connectivity with a finite time budget and a modest requirement on estimation accuracy.

The basis of the trade-off is introduced in the sequel. From the geometric point of view, the smoothing term in $\tilde{\mathscr{R}}(\boldsymbol{\Theta})$ (with $\mu > 0$) increases the sublevel set of $\mathscr{R}(\boldsymbol{\Theta})$, which derives a problem that can be solved via computationally efficient algorithms with an accelerated convergence rate. However, this geometric modification leads to a loss of the estimation accuracy. Thus, it leads to a trade-off between the computational time and estimation accuracy. The trade-off can be identified by Theorem 2.1 based on the statistical dimension of the decent cone of the smoothed regularizer in (2.20), i.e.,

$$\tilde{\mathscr{R}}_G(\tilde{\boldsymbol{\Theta}}) = \mathscr{R}_G(\tilde{\boldsymbol{\Theta}}) + \frac{\mu}{2}\|\tilde{\boldsymbol{\Theta}}\|_F^2. \tag{2.22}$$

We begin with the basic notation used in Proposition 2.1, for some $\tilde{\boldsymbol{\Theta}} \in \mathbb{R}^{2N\times M}$ satisfying $\tilde{\boldsymbol{\Theta}}_{\mathscr{V}_j} = \mathbf{0}$ for $j \neq i$, we have

$$\forall \tilde{\boldsymbol{\Theta}}_{\mathscr{V}_i} \in \mathbb{R}^{2\times M}: \quad \|\tilde{\boldsymbol{\Theta}}_{\mathscr{V}_i}\|_F \geq \|(\tilde{\boldsymbol{\Theta}}_0)_{\mathscr{V}_i}\|_F + \langle \mathbf{Z}_{\mathscr{V}_i}, \tilde{\boldsymbol{\Theta}}_{\mathscr{V}_i} - (\tilde{\boldsymbol{\Theta}}_0)_{\mathscr{V}_i}\rangle, \tag{2.23}$$

which implies $\mathbf{Z}_{\mathscr{V}_i} \in \partial\|(\tilde{\boldsymbol{\Theta}}_0)_{\mathscr{V}_j}\|_F$. In particular, the statistical dimension $\delta(\mathscr{D}(\tilde{\mathscr{R}}_G, \tilde{\boldsymbol{\Theta}}_0))$ can be exactly computed by the following result.

Proposition 2.1 (Statistical Dimension Bound for $\tilde{\mathscr{R}}_G$) *Let* $\boldsymbol{\Theta}_0 \in \mathbb{C}^{N\times M}$ *be with* K *nonzero rows, and define the normalized sparsity as* $\rho := K/N$. *An upper bound of the statistical dimension of the descent cone of* $\tilde{\mathscr{R}}_G$ *at*

$$\tilde{\boldsymbol{\Theta}}_0 = [(\Re\{\boldsymbol{\Theta}_0\})^T, (\Im\{\boldsymbol{\Theta}_0\})^T]^T \in \mathbb{R}^{2N\times M}$$

is given by

$$\frac{\delta(\mathscr{D}(\tilde{\mathscr{R}}_G;\tilde{\boldsymbol{\Theta}}_0))}{N} \leq \inf_{\tau\geq 0}\Big\{\rho(2M+\tau^2(1+2\mu\bar{a}+\mu^2\bar{b})) + (1-\rho)\frac{2^{1-M}}{\Gamma(M)}\int_{\tau}^{\infty}(u-\tau)^2u^{2M-1}e^{-\frac{u^2}{2}}\,\mathrm{d}u\Big\}, \tag{2.24}$$

where $\Gamma(\cdot)$ denotes the Gamma function. The unique optimum $\tau^\star$ which minimizes the right-hand side of (2.24) *is the solution of*

$$\frac{2^{1-M}}{\Gamma(M)}\int_{\tau}^{\infty}\left(\frac{u}{\tau}-1\right)u^{2M-1}e^{-\frac{u^2}{2}}\,\mathrm{d}u = \frac{\rho(1+2\mu\bar{a}+\mu^2\bar{b})}{1-\rho}, \tag{2.25}$$

where $\bar{a}=\frac{1}{S}\sum_{i=1}^{S}\|(\tilde{\boldsymbol{\Theta}}_0)_{\mathscr{V}_i}\|_F$, $\bar{b}=\frac{1}{S}\sum_{i=1}^{S}\|(\tilde{\boldsymbol{\Theta}}_0)_{\mathscr{V}_i}\|_F^2$.

Proof Please refer to Sect. 8.2 for details.

Although the convergence rate of proposed smoothing algorithm, i.e., Algorithm 2.1, can be accelerated by increasing the smoothing parameter, Proposition 2.1 shows that a larger smoothing parameter leads to a larger statistical dimension $\delta(\mathscr{D}(\tilde{\mathscr{R}}_G,\tilde{\boldsymbol{\Theta}}_0))$ since the bound in (2.24) increases with μ.

2.3.3.3 Simulation Results

Proposition 2.1 is verified in Fig. 2.2 with the BS equipped with 2 antennas, the total number of devices being 100, and the channel matrix and pilot matrix generated as

$$\boldsymbol{H}\sim\mathscr{CN}(\boldsymbol{0},\boldsymbol{I}) \quad \text{and} \quad \boldsymbol{Q}\sim\mathscr{CN}(\boldsymbol{0},\boldsymbol{I}),$$

respectively. The recovery is considered to be successful if $\|\hat{\boldsymbol{\Theta}}-\boldsymbol{\Theta}_0\|_F\leq 10^{-5}$. The number of active devices is fixed as $|\mathscr{S}|=10$. Figure 2.2 shows the impact on the exact recovery when changing the smoothing parameter μ. It shows that a larger smoothing parameter will induce a larger statistical dimension of the descent cone of $\tilde{\mathscr{R}}(\boldsymbol{\Theta})$. In other words, longer pilot sequences are required for exact signal recovery.

The effectiveness of the smoothing method illustrated in Algorithm 2.1 is evaluated under the scenario where the base station is equipped with 10 antennas, and the total number of devices is set to be 2000. The number of active devices is fixed as $|\mathscr{S}|=100$. Considering problem (2.15), the channel matrix follows $\boldsymbol{H}\sim\mathscr{CN}(\boldsymbol{0},\boldsymbol{I})$, the pilot matrix follows $\boldsymbol{Q}\sim\mathscr{CN}(\boldsymbol{0},\boldsymbol{I})$ and the additive noise matrix follows $\boldsymbol{N}\sim\mathscr{CN}(\boldsymbol{0},0.01\boldsymbol{I})$. Figure 2.3 demonstrates the convergence rate of Algorithm 2.1 under different smoothing parameters with a fixed pilot sequence length $L=500$. It shows that increasing the smooth parameter enables to accelerate the convergence rate significantly.

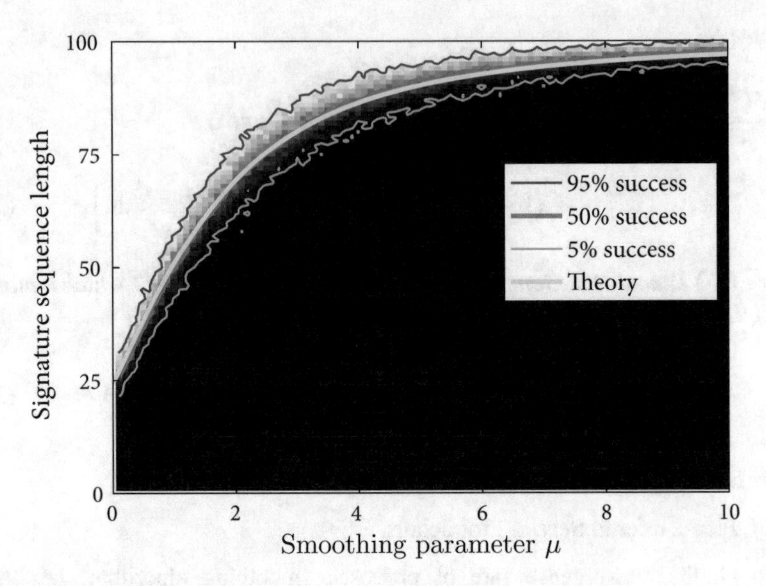

Fig. 2.2 Phase transitions in massive device connectivity via smoothing

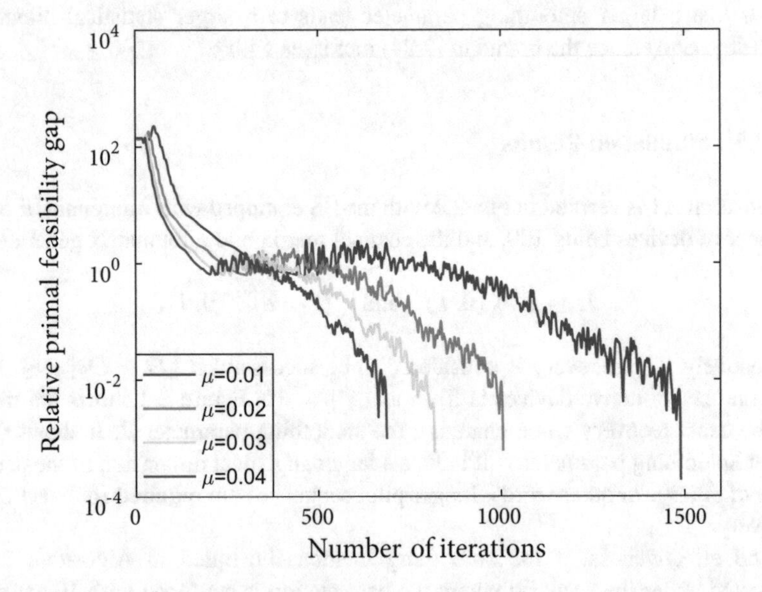

Fig. 2.3 Convergence rate of Algorithm 2.1

Furthermore, with a fixed pilot sequence length $L = 500$, problem (2.15) is solved by Algorithm 2.1 under different smoothing parameters μ. Algorithm 2.1 stops when

$$\left| \|\tilde{\boldsymbol{Q}}\tilde{\boldsymbol{\Theta}} - \tilde{\boldsymbol{Y}}\|_F - \epsilon \right| /\epsilon \leq 10^{-3},$$

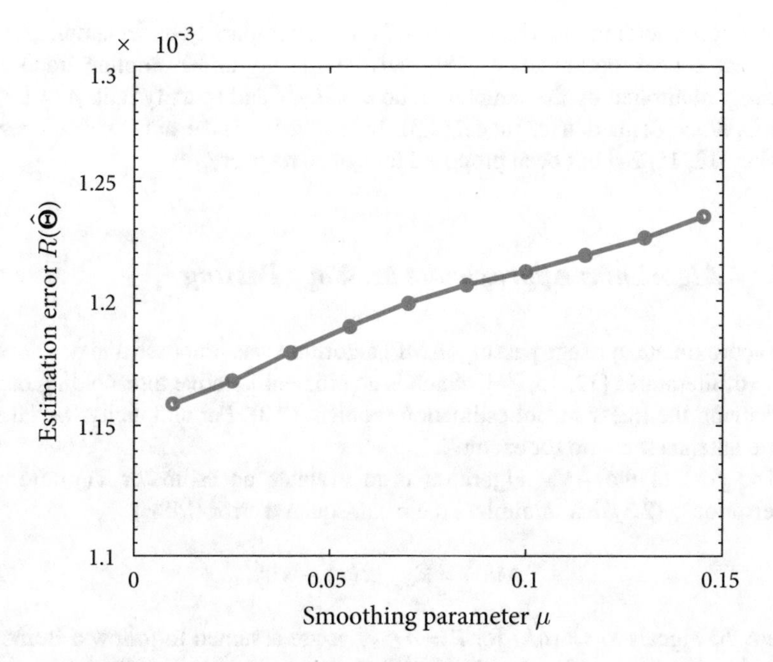

Fig. 2.4 Estimation error versus smoothing parameter μ

where the parameter ϵ is given by

$$\epsilon = \sigma\sqrt{2LM - \delta(\mathscr{D}(\tilde{\mathscr{R}}_G, \tilde{\boldsymbol{\Theta}}_0))}.$$

The simulation result illustrated in Fig. 2.4 is obtained by averaging over 300 channel realizations. It shows that the average squared estimation error becomes large as the smoothing parameter μ increases. This can be justified by Proposition 2.1 that the increase of smoothing parameter results in the increase of statistical dimension $\delta(\mathscr{D}(\tilde{\mathscr{R}}_G, \tilde{\boldsymbol{\Theta}}_0))$.

2.4 Iterative Thresholding Algorithm

Despite attractive theoretical guarantees for the sparse linear model, convex relaxation methods that are solved via a second-order cone program (SOCP) fail in the high-dimensional data setting due to the high computational cost. One way to improve the computational efficiency is the smoothed primal-dual first-order method introduced in the previous section. Another line of literatures that aim to reduce the computational complexity for solving the sparse linear model estimation problem focus on iterative thresholding algorithms [13]. Unfortunately, such fast iterative thresholding algorithms suffer from worse sparsity-undersampling trade-offs than convex optimization [15], and the sparsity-undersampling trade-off is precisely controlled by the sampling ratio $\delta = L/N$ and sparsity ratio $\rho = |\mathscr{S}|/N$ with $L, N, \mathscr{S}$ defined in the model (2.3). To resolve this issue, approximate message passing [12, 15, 24] has been proposed for sparse recovery.

2.4.1 Algorithm: Approximate Message Passing

The approximate message passing (AMP) algorithm was proposed and developed in a line of literatures [12, 15, 24], which is an efficient iterative thresholding method for solving the linear model estimation problem (2.3). For simplicity, we take the single-antenna scenario for example.

The goal of the AMP algorithm is to evaluate an estimator $\hat{\boldsymbol{x}}(\boldsymbol{y})$ from the observation $\boldsymbol{y}$ (2.3) that minimizes the mean-squared error (MSE)

$$\mathrm{MSE} = \mathbb{E}_{\boldsymbol{x}\boldsymbol{y}}||\hat{\boldsymbol{x}}(\boldsymbol{y}) - \boldsymbol{x}||_2^2, \tag{2.26}$$

where the signals $x_i = \alpha_i h_i$ for $i = 1, \ldots, n$ are assumed to follow a Bernoulli–Gaussian distribution. Starting from $\boldsymbol{x}^0 = \boldsymbol{0}$ and $\boldsymbol{r}^0 = \boldsymbol{y}$, the iterative update of the AMP algorithm at the t-th iteration is given by Donoho et al. [15]

$$x_i^{t+1} = \eta_{t,i}((\boldsymbol{r}^t)^{\mathsf{H}}\boldsymbol{a}_i + x_i^t), \tag{2.27a}$$

$$\boldsymbol{r}^{t+1} = \boldsymbol{y} - \boldsymbol{A}\boldsymbol{x}^{t+1} + \frac{N}{L}\boldsymbol{r}^t \sum_{n=1}^{N} \frac{\eta'_{t,i}((\boldsymbol{r}^t)^{\mathsf{H}}\boldsymbol{a}_i + x_i^t)}{N}, \tag{2.27b}$$

where $\boldsymbol{x}^t = [x_1^t, \ldots, x_N^t]^\top \in \mathbb{C}^N$ is the estimate of $\boldsymbol{x}$ at the t-th iteration, $\boldsymbol{r}^t = [r_1^t, \ldots, r_L^t]^\top \in \mathbb{C}^L$ denotes the residual,

$$\eta_{t,i}(\cdot) : \mathbb{C} \to \mathbb{C}$$

is the denoiser which facilitates to induce the sparsity, and $\eta'_{t,i}(\cdot)$ is the first-order derivative of $\eta_{t,i}(\cdot)$. The performance of the AMP algorithm highly depends on the design of the denoiser $\eta_{t,i}(\cdot)$, which will be discussed in the sequel.

2.4.2 Analysis: State Evolution

2.4.2.1 State Evolution

In order to precisely capture the dynamic property of the AMP algorithm, thereby facilitating the design of the denoiser $\eta_{t,i}(\cdot)$, a state evolution formalism was first proposed in the paper [15]. In this formalism, the MSE (2.26) is a state variable and its variation from iteration to iteration can be represented by a plain iterative function, i.e., τ_t.

Define a set of random variables $\hat{X}_i^t$ at the t-th iteration of the AMP algorithm as

$$\hat{X}_i^t = X_i + \tau_t V_i, \quad i = 1, \ldots, n, \tag{2.28}$$

where the distributions of X_i's are characterized by the random variables X_i's, and V_i obeys the normal distribution, i.e., $V_i \in \mathcal{CN}(0, 1)$. In addition, V_i is independent of X_i and V_j for $\forall j \neq i$, and τ_i is the state variable represented as

$$\tau_{t+1}^2 = \frac{\sigma^2}{\xi} + \frac{N}{L}\mathbb{E}\left[|\eta_{t,i}(X_i + \tau_t V_i) - X_i|^2\right], \tag{2.29}$$

where the expectation is over the random variables V_i's and X_i's for $i = 1, \ldots, N$.

The theoretical analysis of AMP is based on the state evolution in the asymptotic regime, when L (i.e., the length of pilot sequences), K (i.e., the average number of active devices in each time slot), N (i.e., the total number of devices) $\to \infty$, while their ratios converge to some positive values $N/L \to \omega$ and

$$K/N \to \epsilon = \lim_{N\to\infty} \sum_i \epsilon_i / N$$

with $\omega, \epsilon \in (0, \infty)$. In massive IoT connectivity, these assumptions imply that the length of the pilot sequence, i.e., L, is in the same order of the number of active users, i.e., K, or total users, i.e., N.

2.4.2.2 Denoiser Designs

In general, the prior distribution of $\boldsymbol{x}$ is assumed to be unknown. In this case, a soft-thresholding denoiser is designed to induce sparsity for $\boldsymbol{x}$, which is given by Donoho et al. [16]:

$$\eta_{t,i}(\hat{x}_i^t) = \left(\hat{x}_i^t - \frac{\theta_i^t \hat{x}_i^t}{|\hat{x}_i^t|}\right)\mathbb{I}(|\hat{x}_i^t| > \theta_i^t), \tag{2.30}$$

where the parameter θ_i^t is the threshold for the i-th device activity detection at the t-th iteration of the AMP algorithm. Based on the state evolution (2.29), the parameter θ_i^t can be optimized to minimize the MSE (2.26). After the t-th iteration proceeded by the AMP algorithm with the denoiser (2.30), device i is evaluated to be active if

$$|(\boldsymbol{r}^t)^{\mathsf{H}}\boldsymbol{a}_i + x_i^t| > \theta_i^t,$$

otherwise it is evaluated to be inactive.

If the prior distribution of $\boldsymbol{x}$ in (2.3) is known, the minimum mean-squared error (MMSE) denoiser via the Bayesian approach can be developed for the AMP algorithm [16]. Based on the random variables defined in (2.28) and assuming the channel signal $h_i \sim \mathcal{CN}(0,1)$ for $i = 1, \ldots, N$, the MMSE denoiser is given in the form of a conditional expectation [16],

$$\begin{aligned}\eta_{t,i}(\hat{x}_i^t) &= \mathbb{E}[X_i|\hat{X}_i^t = \hat{x}_i^t] \\ &= \phi_{t,i}(1+\tau_t)^{-1}\hat{x}_i^t, \quad \forall t, i, \end{aligned} \tag{2.31}$$

where

$$\phi_{t,i} = \frac{1}{1+\frac{1-\epsilon}{\epsilon}\exp\left(-\left(\pi_{t,i}-\psi_{t,i}\right)\right)}, \tag{2.32}$$

$$\pi_{t,i} = (\tau_t^{-2} - (\tau_t^2+1)^{-1})|\hat{x}_i^t|^2, \tag{2.33}$$

$$\psi_{t,i} = \log\det(1+\tau_t^{-2}). \tag{2.34}$$

Note that the above MMSE denoiser is a nonlinear function of $\hat{x}_i^t$ due to the functional form of $\phi_{t,i}$.

2.4.2.3 Asymptotic Performance of Device Activity Detection

Based on the soft thresholding (2.30) and MMSE denoisers (2.31), a miss detection occurs when

$$|(\boldsymbol{r}^t)^{\mathsf{H}}\boldsymbol{a}_i + x_i^t| < \theta_i^t$$

with device i actually being active, while a false alarm occurs when

$$|(\boldsymbol{r}^t)^{\mathsf{H}}\boldsymbol{a}_i + x_i^t| > \theta_i^t$$

with device i actually being inactive. Since the statistical distribution of the thresholding term, i.e., $(\boldsymbol{r}^t)^{\mathsf{H}}\boldsymbol{a}_i + x_i^t$, can be identified by $\hat{x}_i^t$ defined in (2.28), the probabilities of miss detection and false alarm for device i at the t-th iteration of the AMP algorithm can be given by Donoho et al. [15]

$$P_{t,i}^{\mathrm{MD}} = \Pr(\hat{x}_i^t < \theta_i^t | \alpha_i = 1), \tag{2.35}$$

$$P_{t,i}^{\mathrm{FA}} = \Pr(\hat{x}_i^t > \theta_i^t | \alpha_i = 0), \tag{2.36}$$

respectively. The probabilities of missed detection (2.35) and false alarm (2.36) depend on the values of τ_t's (2.29) which can be tracked over iterations based on the state evolution (2.29).

Considering a general multiple-antenna scenario, the theorem in [24] characterizes $P_{t,i}^{\mathrm{MD}}(M)$ and $P_{t,i}^{\mathrm{FA}}$ analytically in terms of τ_t^2 and the number of antennas M. In particular, the miss detection and false alarm probabilities of AMP algorithm with M antennas are denoted by $P_{t,i}^{\mathrm{MD}}(M)$ and $P_{t,i}^{\mathrm{FA}}(M)$. The paper [24] demonstrates that with proper thresholds for device detection, i.e., θ_i^t's, highly accurate device activity detection can be achieved in the asymptotic regime of $M \to \infty$:

$$\lim_{M\to\infty} P_{t,i}^{\mathrm{MD}}(M) = \lim_{M\to\infty} P_{t,i}^{\mathrm{FA}}(M) = 0, \quad \forall t, i. \tag{2.37}$$

It thus indicates that the AMP-based algorithm can accomplish perfect device activity detection in the massive MIMO connectivity systems.

2.4.2.4 Simulation Results

To further illustrate the performance of AMP algorithm for solving the sparse linear model estimation problem, the probabilities of missed detection and false alarm versus the length of the pilot sequences, L, with different numbers of antennas at the BS, i.e., $M = 4, 8$, or 16, are illustrated in Fig. 2.5. In particular, with a given value of M, the average probabilities of missed detection and false alarm over all devices are denoted as

$$P^{\mathrm{MD}}(M) = \sum_{n=1}^{N} P_{\infty,n}^{\mathrm{MD}}(M)/N$$

and

$$P^{\mathrm{FA}}(M) = \sum_{n=1}^{N} P_{\infty,n}^{\mathrm{FA}}(M)/N,$$

respectively, where $P_{\infty,n}^{\mathrm{MD}}(M)$ and $P_{\infty,n}^{\mathrm{FA}}(M)$ are defined in (2.35) and (2.36). Figure 2.5 demonstrates that both P^{MD} and P^{FA} decrease as the pilot sequence length L increases and when M increases.

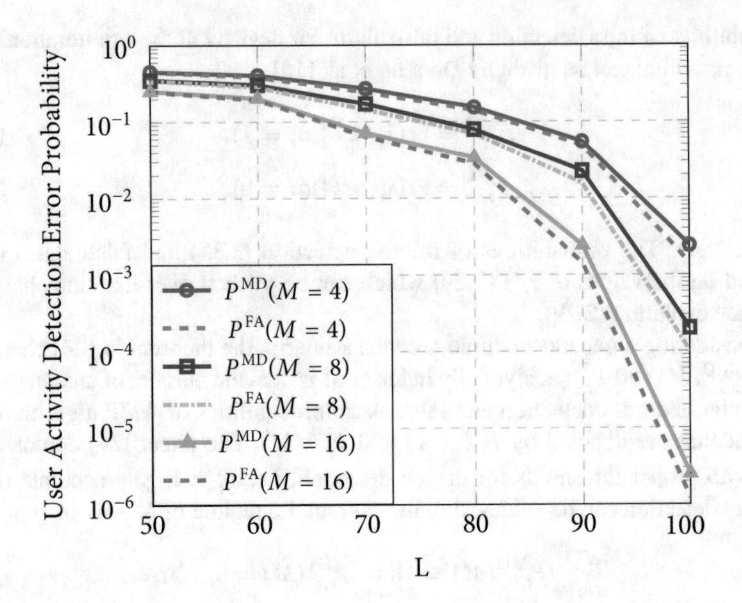

Fig. 2.5 Probabilities of missed detection and false alarm versus pilot sequence lengths

2.5 Summary

This chapter introduced a sparse linear model for joint device activity detection and channel estimation in grant-free random access. Such an access scheme reduces the overhead by removing dedicated channel estimation sequences in IoT networks. To solve the estimation problem, both convex relaxation approach and nonconvex approach have been investigated from the practical and theoretical points of view. Recently, there is a line of studies focusing on solving the sparse linear model via deep-learning-based methods from both empirical and theoretical points of view [5, 11, 17, 20, 37], which provide an also interesting direction for future study.

References

1. Amelunxen, D., Lotz, M., McCoy, M.B., Tropp, J.A.: Living on the edge: phase transitions in convex programs with random data. Inf. Inference **3**(3), 224–294 (2014)
2. Bastug, E., Bennis, M., Debbah, M.: Living on the edge: the role of proactive caching in 5G wireless networks. IEEE Commun. Mag. **52**(8), 82–89 (2014)
3. Bayati, M., Montanari, A.: The dynamics of message passing on dense graphs, with applications to compressed sensing. IEEE Trans. Inf. Theory **57**(2), 764–785 (2011)
4. Becker, S.R., Candès, E.J., Grant, M.C.: Templates for convex cone problems with applications to sparse signal recovery. Math. Program. Comput. **3**(3), 165–218 (2011)
5. Borgerding, M., Schniter, P., Rangan, S.: AMP-inspired deep networks for sparse linear inverse problems. IEEE Trans. Signal Process. **65**(16), 4293–4308 (2017)

6. Boyd, S., Parikh, N., Chu, E., Peleato, B., Eckstein, J.: Distributed optimization and statistical learning via the alternating direction method of multipliers. Found. Trends Mach. Learn. **3**(1), 1–122 (2011)
7. Bruer, J.J., Tropp, J.A., Cevher, V., Becker, S.R.: Designing statistical estimators that balance sample size, risk, and computational cost. IEEE J. Sel. Top. Sign. Proces. **9**(4), 612–624 (2015)
8. Candes, E., Tao, T.: Near optimal signal recovery from random projections: universal encoding strategies? IEEE Trans. Inf. Theory **52**(12), 5406–5425 (2006)
9. Chandrasekaran, V., Jordan, M.I.: Computational and statistical tradeoffs via convex relaxation. Proc. Natl. Acad. Sci. **110**(13), E1181–E1190 (2013)
10. Chen, S.S., Donoho, D.L., Saunders, M.A.: Atomic decomposition by basis pursuit. SIAM Rev. **43**(1), 129–159 (2001)
11. Chen, X., Liu, J., Wang, Z., Yin, W.: Theoretical linear convergence of unfolded ISTA and its practical weights and thresholds. In: Advances in Neural Information Processing Systems (NeurIPS), pp. 9061–9071 (2018)
12. Chen, Z., Sohrabi, F., Yu, W.: Sparse activity detection for massive connectivity. IEEE Trans. Signal Process. **66**(7), 1890–1904 (2018)
13. Daubechies, I., Defrise, M., De Mol, C.: An iterative thresholding algorithm for linear inverse problems with a sparsity constraint. Commun. Pure Appl. Math. **57**(11), 1413–1457 (2004)
14. Donoho, D.L., et al.: Compressed sensing. IEEE Trans. Inf. Theory **52**(4), 1289–1306 (2006)
15. Donoho, D.L., Maleki, A., Montanari, A.: Message-passing algorithms for compressed sensing. Proc. Nat. Acad. Sci. **106**(45), 18914–18919 (2009)
16. Donoho, D.L., Maleki, A., Montanari, A.: Message passing algorithms for compressed sensing: I. motivation and construction. In: Proceedings of the IEEE Information Theory Workshop on Information Theory, pp. 1–5. IEEE, Piscataway (2010)
17. Fletcher, A.K., Pandit, P., Rangan, S., Sarkar, S., Schniter, P.: Plug-in estimation in high-dimensional linear inverse problems: a rigorous analysis. In: Advances in Neural Information Processing Systems (NeurIPS), pp. 7440–7449 (2018)
18. Giryes, R., Eldar, Y.C., Bronstein, A.M., Sapiro, G.: Tradeoffs between convergence speed and reconstruction accuracy in inverse problems. IEEE Trans. Signal Process. **66**(7), 1676–1690 (2018)
19. Goldstein, T., O'Donoghue, B., Setzer, S., Baraniuk, R.: Fast alternating direction optimization methods. SIAM J. Imaging Sci. **7**(3), 1588–1623 (2014)
20. Ito, D., Takabe, S., Wadayama, T.: Trainable ISTA for sparse signal recovery. IEEE Trans. Signal Process. **67**(12), 3113–3125 (2019)
21. Jiang, T., Shi, Y., Zhang, J., Letaief, K.B.: Joint activity detection and channel estimation for IoT networks: phase transition and computation-estimation tradeoff. IEEE Internet Things J. **6**(4), 6212–6225 (2018)
22. Lai, M.J., Yin, W.: Augmented ℓ_1 and nuclear-norm models with a globally linearly convergent algorithm. SIAM J. Imaging Sci. **6**(2), 1059–1091 (2013)
23. Lan, G., Lu, Z., Monteiro, R.D.: Primal-dual first-order methods with $\mathcal{O}(1/\epsilon)$ iteration-complexity for cone programming. Math. Program. **126**(1), 1–29 (2011)
24. Liu, L., Yu, W.: Massive connectivity with massive MIMO–part I: device activity detection and channel estimation. IEEE Trans. Signal Process. **66**(11), 2933–2946 (2018)
25. Liu, L., Yu, W.: Massive connectivity with massive MIMO—part II: achievable rate characterization. IEEE Trans. Signal Process. **66**(11), 2947–2959 (2018)
26. Liu, L., Larsson, E.G., Yu, W., Popovski, P., Stefanovic, C., De Carvalho, E.: Sparse signal processing for grant-free massive connectivity: a future paradigm for random access protocols in the Internet of Things. IEEE Signal Process. Mag. **35**(5), 88–99 (2018)
27. Muthukrishnan, S., et al.: Data streams: algorithms and applications. Found. Trends Theor. Comput. Sci. **1**(2), 117–236 (2005)
28. Oymak, S., Hassibi, B.: Sharp MSE bounds for proximal denoising. Found. Comput. Math. **16**(4), 965–1029 (2016)
29. Oymak, S., Recht, B., Soltanolkotabi, M.: Sharp time–data tradeoffs for linear inverse problems. IEEE Trans. Inf. Theory **64**(6), 4129–4158 (2018)

30. Parikh, N., Boyd, S.: Proximal algorithms. Found. Trends Optim. **1**(3), 127–239 (2014)
31. Schepker, H.F., Bockelmann, C., Dekorsy, A.: Exploiting sparsity in channel and data estimation for sporadic multi-user communication. In: Proceedings of the International Symposium on Wireless Communication Systems, pp. 1–5. VDE, Frankfurt (2013)
32. Schneider, R., Weil, W.: Stochastic and Integral Geometry. Springer, Berlin (2008)
33. Shi, Y., Zhang, J., Letaief, K.B., Bai, B., Chen, W.: Large-scale convex optimization for ultra-dense Cloud-RAN. IEEE Trans. Wirel. Commun. **22**(3), 84–91 (2015)
34. Wunder, G., Boche, H., Strohmer, T., Jung, P.: Sparse signal processing concepts for efficient 5G system design. IEEE Access **3**, 195–208 (2015)
35. Wunder, G., Jung, P., Wang, C.: Compressive random access for post-LTE systems. In: Proceedings of the IEEE International Conference on Communications Workshops (ICC) Workshops, pp. 539–544. IEEE, Piscataway (2014)
36. Xu, X., Rao, X., Lau, V.K.: Active user detection and channel estimation in uplink CRAN systems. In: Proceedings of the International Conference on Communications (ICC), pp. 2727–2732. IEEE, Piscataway (2015)
37. Zhang, J., Ghanem, B.: ISTA-net: Interpretable optimization-inspired deep network for image compressive sensing. In: Proceedings of the IEEE Conference on Computer Vision and Pattern Recognition (CVPR), pp. 1828–1837 (2018)

Chapter 3
Blind Demixing

Abstract This chapter presents a blind demixing model for joint data decoding and channel estimation in IoT networks, without transmitting pilot sequences. The problem formulation based on the cyclic convolution in the time domain is first introduced, which is then reformulated in the Fourier domain for the ease of algorithm design. A convex relaxation approach based on nuclear norm minimization is first presented as a basic solution. Next, several nonconvex approaches are introduced, including both regularized and regularization-free Wirtinger flow and the Riemannian optimization algorithm. The mathematical tools for analyzing nonconvex approaches are also provided.

3.1 Joint Data Decoding and Channel Estimation

For data transmission in IoT networks, as the blocklength of packets is typically very short, the channel estimation sequences (CES) (illustrated in Fig. 1.1) occupy the primary part of the packet [11]. Thus, CES overhead reduction becomes critical to achieve low-overhead communications. To exclude the CES overhead, the BS may jointly decode data and estimate channel states, which can be established as a blind demixing model (3.1) [8].

For an IoT network containing one BS and s devices, as shown in Fig. 3.1, the observation signal vector is the mixture of the encoding signals generated from s devices and passed through the corresponding channels. The goal of the BS is to jointly decode data and estimate the channel states, which can be captured via a *blind demixing model* consisting of both summation operation and convolution operation. For ease of algorithm design, the measurements in the blind demixing are represented in the Fourier domain, which are given by

$$y_j = \sum_{i=1}^{s} \boldsymbol{b}_j^{\mathsf{H}} \boldsymbol{h}_i \boldsymbol{x}_i^{\mathsf{H}} \boldsymbol{a}_{ij}, \ 1 \le j \le L. \tag{3.1}$$

Y. Shi et al., *Low-overhead Communications in IoT Networks*,
https://doi.org/10.1007/978-981-15-3870-4_3

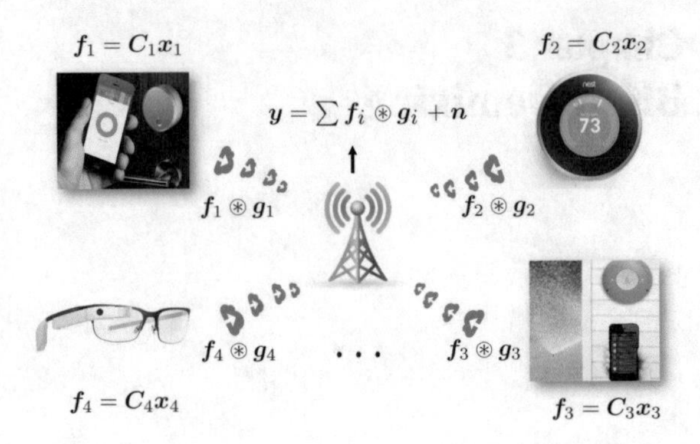

Fig. 3.1 The blind demixing model in an IoT network

Denote

$$\boldsymbol{y} = [y_1, \ldots, y_L]^\top \in \mathbb{C}^L$$

as the received signal at the BS represented in the Fourier domain, $\{\boldsymbol{b}_j\}$, $\{\boldsymbol{a}_{ij}\}$ are design vectors, and $\{\boldsymbol{h}_i\}$, $\{\boldsymbol{x}_i\}$ are channel states and data signals, respectively. Particularly, the design vectors $\{\boldsymbol{b}_j\}$ indicate the Fourier transform operation and the design vectors $\{\boldsymbol{a}_{ij}\}$ indicate the encoding procedure. By evaluating the vectors $\{\boldsymbol{h}_i\}$, $\{\boldsymbol{x}_i\}$ from the observation $\boldsymbol{y}$, data decoding and channel estimation can be simultaneously accomplished.

There is a growing body of recent works paying attention on the blind demixing model (3.1). In particular, semidefinite programming has been developed in [9] to solve the blind demixing problems by lifting the bilinear model into the rank-one matrix model. However, it is computationally expensive to deal with large-scale problems. To address this issue, nonconvex algorithms, e.g., regularized Wirtinger flow with spectral initialization [10], have been developed to optimize the variables in the vector space. The Riemannian trust-region optimization algorithm without regularization was further developed in [8] to improve the convergence rate compared to the regularized Wirtinger flow algorithm [10]. Recently, concerning the blind demixing problem, theoretical guarantees for regularization-free Wirtinger flow with spectral initialization were established in [6]. To further find a natural initialization for the practitioners that works equally well as spectral initialization, the paper [7] established the global convergence guarantee of Wirtinger flow with random initialization for blind demixing.

In the sequel, the procedure of establishing the blind demixing model based on the convolution operations in IoT networks will be first illustrated. We further clarify the vital role that the blind demixing model plays in joint data decoding and channel estimation. Then, effective algorithms and rigorous analysis are provided for both convex and nonconvex approaches.

3.2 Problem Formulation

In this section, the basic concept of the cyclic convolution is first introduced, followed by a detailed description of the blind demixing model based on the cyclic convolution.

3.2.1 *Cyclic Convolution*

The elementary concept of the cyclic convolution is first introduced to characterize the connection among the channel state, received signal, and transmitted signal, thereby assisting the presentation of the blind demixing model.

Denote $p[n]$ and $\theta[n]$ as the transmitted signal and received signal in the n-th time slot, respectively. Define q_ℓ as the ℓ-th tap channel impulse response which is constant with n. Hence, the channel is assumed to be linear time-invariant. Thus, the discrete-time model is represented as

$$\theta[n] = \sum_{\ell=0}^{L_t-1} q_\ell p[n-\ell], \tag{3.2}$$

where L_t is the number of nonzero taps. A *cyclic prefix* is added to $\boldsymbol{p}$, which yields the symbol vector, i.e., $\boldsymbol{d} \in \mathbb{C}^{N_p+L_t-1}$:

$$\boldsymbol{d} = [p[N_p - L_t + 1], \ldots, p[N_p - 1], p[0], p[1], \ldots, p[N_p - 1]]^\top. \tag{3.3}$$

The output over the time interval $n \in [L_t, N_p + L_t - 1]$ is represented as

$$\theta[n] = \sum_{\ell=0}^{L_t-1} q_\ell d[(n - L_t - \ell) \text{ modulo } N_p]. \tag{3.4}$$

Denote the output of length N_p as

$$\boldsymbol{\theta} = [\theta[L_t], \ldots, \theta[N_p + L_t - 1]]^\top, \tag{3.5}$$

and the channel impulse as

$$\boldsymbol{q} = [q_0, q_1, \ldots, q_{L_t-1}, 0, \ldots, 0]^\top \in \mathbb{C}^{N_p},$$

and then (3.4) can be reformulated as

$$\boldsymbol{\theta} = \boldsymbol{q} \circledast \boldsymbol{p},$$

where the notion $\circledast$ denotes the *cyclic convolution*.

3.2.2 System Model

Consider a network with one BS and s devices. Denote the original signals of length N from the i-th user as $\boldsymbol{x}_i \in \mathbb{C}^N$. The transmitted signals over L time slots from the i-th user are represented as

$$\boldsymbol{f}_i = \boldsymbol{C}_i \boldsymbol{x}_i, \tag{3.6}$$

where $\boldsymbol{C}_i \in \mathbb{C}^{L \times N}$ with $L > N$ as the encoding matrix and known to the BS. The signals $\boldsymbol{f}_i$'s are transmitted through individual time-invariant channels endowed with impulse responses $\boldsymbol{h}_i$'s where a maximum delay of at most K samples is contained in $\boldsymbol{h}_i \in \mathbb{C}^K$. The zero-padded channel vector $\boldsymbol{g}_i \in \mathbb{C}^L$ is given as

$$\boldsymbol{g}_i = [\boldsymbol{h}_i^\top, 0, \ldots, 0]^\top. \tag{3.7}$$

Hence, based on the cyclic convolution operation, the received signal is given as

$$\boldsymbol{z} = \sum\nolimits_{i=1}^{s} \boldsymbol{f}_i \circledast \boldsymbol{g}_i + \boldsymbol{n}, \tag{3.8}$$

where $\boldsymbol{n}$ is the additive white complex Gaussian noise. The BS needs to recover the data signals $\{\boldsymbol{x}_i\}_{i=1}^s$ from the observation $\boldsymbol{z}$ without knowing channel states $\{\boldsymbol{g}_i\}_{i=1}^s$. This model is called a *blind demixing* model.

3.2.3 Representation in the Fourier Domain

For the ease of algorithm design and theoretical analysis, the blind demixing model based on cyclic convolution is represented in the Fourier domain. This is achieved by left multiplying the signals in the time domain with the unitary discrete Fourier transform (DFT) matrix and converting the convolution operation in the time domain to the componentwise production operation in the Fourier domain [8, 9]:

$$\boldsymbol{y} = \boldsymbol{F}\boldsymbol{z} = \sum_i (\boldsymbol{F}\boldsymbol{C}_i\boldsymbol{x}_i) \odot \boldsymbol{B}\boldsymbol{h}_i + \boldsymbol{F}\boldsymbol{n}, \tag{3.9}$$

where the operation $\odot$ is the componentwise product. Here, the first K columns of the unitary discrete Fourier transform (DFT) matrix $\boldsymbol{F} \in \mathbb{C}^{L \times L}$ with $\boldsymbol{F}\boldsymbol{F}^{\mathsf{H}} = \boldsymbol{I}_L$ form the known matrix

$$\boldsymbol{B} := [\boldsymbol{b}_1, \ldots, \boldsymbol{b}_L]^{\mathsf{H}} \in \mathbb{C}^{L \times K}$$

with $\boldsymbol{b}_j \in \mathbb{C}^K$ for $1 \le j \le L$. An example of the sparse linear model is illustrated in Example 3.1.

Example 3.1 Consider a network with two devices and one single-antenna BS. We assume that $K = N = 1$ and the data signals $\{\boldsymbol{x}_i\}_{i=1}^2$ as $\boldsymbol{x}_1 = 1, \boldsymbol{x}_2 = 2$, and channel signals $\{\boldsymbol{h}_i\}_{i=1}^2$ as $\boldsymbol{h}_1 = 3, \boldsymbol{h}_2 = 4$. In addition, three time slots are considered in this example such that the encoding matrices are given by

$$\boldsymbol{C}_1 = \begin{bmatrix} 1 \\ 1 \\ 2 \end{bmatrix} \quad \text{and} \quad \boldsymbol{C}_2 = \begin{bmatrix} 4 \\ 2 \\ 1 \end{bmatrix}. \tag{3.10}$$

Based on the unitary discrete Fourier transform (DFT) matrix $\boldsymbol{F} \in \mathbb{C}^{3\times 3}$:

$$\boldsymbol{F} = \begin{bmatrix} 0.5774 & 0.5774 & 0.5774 \\ 0.5774 & -0.2887 - 0.5i & -0.2887 + 0.5i \\ 0.5774 & -0.2887 + 0.5i & -0.2887 - 0.5i \end{bmatrix}, \tag{3.11}$$

it yields the blind demixing model:

$$\begin{aligned} \boldsymbol{y} &= \sum_i (\boldsymbol{F}\boldsymbol{C}_i\boldsymbol{x}_i) \odot \boldsymbol{B}\boldsymbol{h}_i \\ &= \begin{bmatrix} 2.3096 \\ -0.2887 + 0.5i \\ -0.2887 - 0.5i \end{bmatrix} \odot \begin{bmatrix} 1.7322 \\ 1.7322 \\ 1.7322 \end{bmatrix} + \begin{bmatrix} 8.8036 \\ 2.8870 - 0.5i \\ 2.8870 + 0.5i \end{bmatrix} \odot \begin{bmatrix} 2.3096 \\ 2.3096 \\ 2.3096 \end{bmatrix} \\ &= \begin{bmatrix} 22.6706 \\ 6.1677 - 1.4435i \\ 6.1677 + 1.4435i \end{bmatrix}. \end{aligned} \tag{3.12}$$

Generally, the blind demixing model can be formulated as the sum of bilinear measurements of vectors $\boldsymbol{x}_i^\natural \in \mathbb{C}^N, \boldsymbol{h}_i^\natural \in \mathbb{C}^K, i = 1, \ldots, s$, i.e.,

$$y_j = \sum_{i=1}^{s} \boldsymbol{b}_j^{\mathsf{H}} \boldsymbol{h}_i^\natural \boldsymbol{x}_i^{\natural\mathsf{H}} \boldsymbol{a}_{ij} + e_j,\ 1 \le j \le L, \tag{3.13}$$

where y_j is the j-th entry of $\boldsymbol{y}$ in (3.9), $\boldsymbol{b}_j \in \mathbb{C}^K$ denotes the j-th column of $\boldsymbol{B}^{\mathsf{H}}$, and $\boldsymbol{a}_{ij} \in \mathbb{C}^N$ denotes the j-th column of $(\boldsymbol{F}\boldsymbol{C}_i)^{\mathsf{H}}$. Furthermore, the noise e_j obeys

$$e_j \sim \mathscr{CN}\left(0, \frac{\sigma^2 d_0^2}{2L}\right) \tag{3.14}$$

with

$$d_0 = \sqrt{\sum_{i=1}^{s} \|\boldsymbol{h}_i^{\natural}\|_2^2 \|\boldsymbol{x}_i^{\natural}\|_2^2} \tag{3.15}$$

and σ^2 as the measurement of noise variance.

3.3 Convex Relaxation Approach

In this section, a convex relaxation approach for estimating the blind demixing model is introduced, followed by theoretical analysis.

3.3.1 Method: Nuclear Norm Minimization

To begin with, a low-rank matrix optimization problem is established via lifting the bilinear model in (3.13) to the rank-one matrices space. Based on (3.9), the j-th entry of the first term in (3.9) can be formulated as

$$[(\boldsymbol{F}\boldsymbol{C}_i\boldsymbol{x}_i) \odot \boldsymbol{B}\boldsymbol{h}_i]_j = (\boldsymbol{c}_{ij}^{\mathsf{H}}\boldsymbol{x}_i)(\boldsymbol{b}_j^{\mathsf{H}}\boldsymbol{h}_i) = \left\langle \boldsymbol{c}_{ij}\bar{\boldsymbol{b}}_j^{\mathsf{H}}, \boldsymbol{X}_i \right\rangle,$$

where $\boldsymbol{c}_{ij}^{\mathsf{H}}$ is the j-th row of $\boldsymbol{F}\boldsymbol{C}_i$, $\boldsymbol{b}_j^{\mathsf{H}}$ is the j-th row of $\boldsymbol{B}$, and $\boldsymbol{X}_i = \boldsymbol{x}_i\bar{\boldsymbol{h}}_i^{\mathsf{H}} \in \mathbb{C}^{N\times K}$ is a rank-one matrix. We have

$$y_j = \left\langle \begin{bmatrix} \boldsymbol{x}_1\bar{\boldsymbol{h}}_1^{\mathsf{H}} & 0 & \cdots & 0 \\ 0 & \boldsymbol{x}_2\bar{\boldsymbol{h}}_2^{\mathsf{H}} & \cdots & 0 \\ \vdots & \vdots & \ddots & \vdots \\ 0 & 0 & \cdots & \boldsymbol{x}_s\bar{\boldsymbol{h}}_s^{\mathsf{H}} \end{bmatrix}, \begin{bmatrix} \boldsymbol{c}_{1j}\bar{\boldsymbol{b}}_j^{\mathsf{H}} & 0 & \cdots & 0 \\ 0 & \boldsymbol{c}_{2j}\bar{\boldsymbol{b}}_j^{\mathsf{H}} & \cdots & 0 \\ \vdots & \vdots & \ddots & \vdots \\ 0 & 0 & \cdots & \boldsymbol{c}_{sj}\bar{\boldsymbol{b}}_j^{\mathsf{H}} \end{bmatrix} \right\rangle + e_l. \tag{3.16}$$

Thus, the received signal at the BS in the Fourier domain is given by

$$\boldsymbol{y} = \sum\nolimits_{k=1}^{s} \mathscr{A}_i(\boldsymbol{X}_i) + \boldsymbol{e}, \tag{3.17}$$

where the vector $\boldsymbol{e}$ denotes the additive Gaussian noise and the linear operator $\mathscr{A}_i$: $\mathbb{C}^{N\times K} \to \mathbb{C}^L$ is represented as

$$\mathscr{A}_i(\boldsymbol{X}_i) := \left\{ \left\langle \boldsymbol{c}_{ij}\bar{\boldsymbol{b}}_j^{\mathsf{H}}, \boldsymbol{X}_i \right\rangle \right\}_{i=1}^{L} = \{\langle \boldsymbol{A}_{ij}, \boldsymbol{X}_i \rangle\}_{i=1}^{L}, \tag{3.18}$$

with

$$A_{ij} = c_{ij}\bar{b}_j^{\mathsf{H}}.$$

In addition, the operation $\mathscr{A}_i^* : \mathbb{C}^L \rightarrow \mathbb{C}^{N\times K}$ can be represented as

$$\mathscr{A}_i^*(y) = \sum_{j=1}^{L} y_j b_j c_{ij}^{\mathsf{H}}. \tag{3.19}$$

The goal of blind demixing problem is to find the rank-one matrices that match the observation, formulated as

$$\begin{aligned} &\text{find } \operatorname{rank}(W_i) = 1,\ i = 1, \ldots, s \\ &\text{subject to } \left\| \sum_{i=1}^{s} \mathscr{A}_i(W_i) - y \right\|_2 \leq \varepsilon, \end{aligned} \tag{3.20}$$

where the parameter ε is a bound for $\|e\|_2$ (recall that e appeared in (3.17)). Nevertheless, due to the nonconvexity of the rank function, problem (3.20) is NP-hard and thus intractable. The nuclear norm minimization approach has been proposed to relax the rank function [9], which gives the following formulation:

$$\begin{aligned} &\underset{W_i, i=1,\ldots,s}{\text{minimize}} \sum_{i=1}^{s} \| W_i \|_* \\ &\text{subject to } \left\| \sum_{i=1}^{s} \mathscr{A}_i(W_i) - y \right\|_2 \leq \varepsilon. \end{aligned} \tag{3.21}$$

The problem (3.21) can be solved via semidefinite program. Based on the estimated $\hat{X}_i$, the corresponding $\hat{h}_i$ and $\hat{x}_i$ can be set as the right and left singular values of $\hat{X}_i$, respectively.

3.3.2 Theoretical Analysis

To present theoretical analysis for methods that solve the blind demixing problem, several notions are first introduced. For simplicity, we summarize the parameters involved in the analysis of solving the blind demixing problem (3.21) via semidefinite program in Table 3.1, and the detailed formulations of these parameters can be found in [9].

The paper [9] demonstrates that the method (3.21) provides an effective way to solve the blind demixing problem and is also robust to noise, as illustrated in Theorem 3.1.

Table 3.1 Conditions involved in Theorem 3.1 and corresponding section mentioned in [9]

Condition	Parameter	Reference
Joint incoherent pattern on the matrices $\boldsymbol{B}$	$\mu_{\max}, \mu_{\min}$	Sect. II-C
Incoherence between $\boldsymbol{b}_j$ and $\boldsymbol{h}_i$	μ	Sect. II-D
Upper bound on $\\|\mathscr{A}_i\\| := \sup_{\boldsymbol{X}\neq\boldsymbol{0}} \\|\mathscr{A}_i(\boldsymbol{X})\\|_F / \\|\boldsymbol{X}\\|_F$	γ	Sect. II-E

Theorem 3.1 *Considering the blind demixing model (3.17) in the noiseless scenario, if*

$$L \geq Cs^2 \max\{\mu_{\max}^2 K, \mu_h^2 N\} \log^2 L \log \gamma,$$

where $C > 0$ is sufficiently large, and $\mu_{\max}, \mu, \gamma$ are summarized in Table 3.1, then the convex relaxation approach (3.21) recovers the ground truth rank-one matrices exactly with high probability.

Proof The proof details of Theorem 3.1 can be referred to the paper [9] and the proof architecture of Theorem 3.1 is briefly summarized. We further present a sufficient condition and an approximate dual certificate condition for the minimizer of (3.21) to be the unique solution to (3.20). These conditions stipulate that matrices $\mathscr{A}_i$ need to satisfy two key properties. The first property can be regarded as a modification of the celebrated *Restricted Isometry Property (RIP)* [3], as it requires $\mathscr{A}_i$ to act in a certain sense as "local" approximate isometries [4]. The second property requires the two operators $\mathscr{A}_i$ and $\mathscr{A}_j$ to satisfy a *"local" mutual incoherence property*. With these two key properties in place, an approximate dual certificate can be established that fulfills the sufficient condition. With all these tools in place, the proof of Theorem 3.1 can be completed.

Remark 3.1 Theorem 3.1 demonstrates that the successful recovery of $\{\boldsymbol{W}_i\}_{i=1}^s$ in problem (3.21) in the noiseless scenario via semidefinite programming can be achieved with high probability as long as the number of measurements satisfies

$$L \gtrsim s^2 \max\{\mu_{\max}^2 K, \mu_h^2 N\} \log^2 L.$$

The paper [9] further considers problem (3.21) in the noisy scenario and provides the performance guarantee of recovering $\{\boldsymbol{W}_i\}_{i=1}^s$ in problem (3.21) under the same conditions as in Theorem 3.1.

3.4 Nonconvex Approaches

While convex techniques can be exploited to solve the blind demixing problem provably and robustly under certain assumptions, the resulting algorithms are computationally expensive for large-scale problems. This motivates the development of efficient nonconvex approaches, which are introduced in this section. The nonconvex approaches introduced in this section can be separated into two types: the Wirtinger flow based approach, which is an iterative algorithm based on the gradients derived in the complex space, and the Riemannian optimization based approach, which is developed on the Riemannian manifold search space.

3.4.1 Regularized Wirtinger Flow

Considering the rank-one structure of the blind demixing model, matrix factorization provides an efficient method to address the low-rank optimization problem. Specifically, Ling and Strohmer [10] developed an algorithm to solve the blind demixing problem based on matrix factorization and the regularized Wirtinger flow. The regularized optimization problem is established as

$$\underset{\boldsymbol{u}_k,\boldsymbol{v}_k,k=1,\ldots,s}{\text{minimize}} \quad F(\boldsymbol{u},\boldsymbol{v}) := g(\boldsymbol{u},\boldsymbol{v}) + \lambda R(\boldsymbol{u},\boldsymbol{v}), \tag{3.22}$$

where

$$g(\boldsymbol{u},\boldsymbol{v}) := \left\| \sum\nolimits_{k=1}^{s} \mathscr{A}_k\big(\boldsymbol{u}_k\boldsymbol{v}_k^{\mathsf{H}}\big) - \boldsymbol{y} \right\|^2$$

with $\boldsymbol{u}_k \in \mathbb{C}^N$, $\boldsymbol{v}_k \in \mathbb{C}^K$ and the aim of the regularizer $R(\boldsymbol{u},\boldsymbol{v})$ is to enable the iterates to lie in the *basin of attraction* [10]. The algorithm begins with a spectral initialization point and updates the iterates as:

$$\boldsymbol{u}_k^{[t+1]} = \boldsymbol{u}_k^{[t]} - \eta\nabla F_{\boldsymbol{u}_k}\big(\boldsymbol{u}_k^{[t]}, \boldsymbol{v}_k^{[t]}\big), \tag{3.23}$$

$$\boldsymbol{v}_k^{[t+1]} = \boldsymbol{v}_k^{[t]} - \eta\nabla F_{\boldsymbol{v}_k}\big(\boldsymbol{u}_k^{[t]}, \boldsymbol{v}_k^{[t]}\big), \tag{3.24}$$

where $\nabla F_{\boldsymbol{u}_k}$ is the derivative of the objective function (3.22) with respect to $\boldsymbol{u}_k$. The following theorem provided in [10] demonstrates that the regularized Wirtinger gradient descent will guarantee the linear convergence of the iterates, and the recovery is exact in the noiseless scenario and stable in the presence of noise.

Denote the condition number as

$$\kappa := \frac{\max_i \|\boldsymbol{x}_i^{\natural}\|_2}{\min_i \|\boldsymbol{x}_i^{\natural}\|_2}, \tag{3.25}$$

and recall that d_0 in (3.15). Furthermore, for simplicity, we summarize the parameters involved in the analysis of solving the blind demixing problem (3.22) via regularized Wirtinger flow in Table 3.2, and the detailed formulations of these parameters can be referred to the references presented in the table.

Table 3.2 Conditions involved in Theorem 3.2 and corresponding section mentioned in [10]

Condition	Parameter	Reference
Local regularity condition	ω	Sect. 5.1
Robustness condition on $\|\mathscr{A}^*(\boldsymbol{e})\|$	γ_e	Sect. 6.5

Algorithm 3.1: Initialization via spectral method and projection

1: for $i = 1, 2, \ldots, s$ do
2: Compute $\mathscr{A}_i^*(\boldsymbol{y})$.
3: Find the leading singular value, left and right singular vectors of $\mathscr{A}_i^*(\boldsymbol{y})$, denoted by $(d_i, \hat{\boldsymbol{h}}_{i0}, \hat{\boldsymbol{x}}_{i0})$.
4: Solve the following optimization problem for $1 \leq i \leq s$:

$$\mu_i^{(0)} := \text{argmin}_{z \in \mathbb{C}^K} \|z - \sqrt{d_i}\hat{\boldsymbol{h}}_{i0}\|^2 \text{ s.t. } \sqrt{L}\|\boldsymbol{B}z\|_\infty \leq 2\sqrt{d_i}\mu.$$

5: Set $\boldsymbol{v}_i^{(0)} = \sqrt{d_i}\hat{\boldsymbol{x}}_{i0}$.
6: end for
7: Output: $\{(\boldsymbol{u}_i^{(0)}, \boldsymbol{v}_i^{(0)}, d_i)\}_{i=1}^s$ or $(\boldsymbol{u}^{(0)}, \boldsymbol{v}^{(0)}, \{d_i\}_{i=1}^s)$.

Theorem 3.2 *Starting from the initial point generated via Algorithm 3.1, the regularized Wirtinger flow algorithm derives a sequence of iterates* $(\boldsymbol{u}^{[t]}, \boldsymbol{v}^{[t]})$ *which converges to the global minimum linearly,*

$$\sum_{k=1}^{s} \left\| \boldsymbol{u}_k^{[t]}(\boldsymbol{v}_k^{[t]})^{\mathsf{H}} - \boldsymbol{h}_k^{\natural}\boldsymbol{x}_k^{\natural\mathsf{H}} \right\|_F \leq \frac{d_0}{\sqrt{2s\kappa^2}}(1 - \eta\omega)^{t/2} + 60\sqrt{s}\gamma_e \tag{3.26}$$

with high probability if the number of measurements L satisfies

$$L \geq C(\mu^2 + \sigma^2)s^2\kappa^4 \max\{K, N\} \log^2 L, \tag{3.27}$$

where $C > 0$ *is sufficiently large. Here, the parameter* σ *and* d_0 *are defined in (3.14), and* ω, γ_e *are summarized in Table 3.2.*

Proof The convergence analysis provided in Theorem 3.2 relies on four conditions: local regularity condition of the objective function $F(\boldsymbol{u}, \boldsymbol{v})$ (3.22), local smoothness condition of the objective function $F(\boldsymbol{u}, \boldsymbol{v})$ (3.22), local restricted isometry property, and robustness condition. Under the assumptions mentioned in Theorem 3.2, with the spectral initialization being in the basin of attraction, the four conditions can be guaranteed, which yield the results of Theorem 3.2.

The performance of regularized Wirtinger flow is further illustrated in Sect. 3.4.4.

Remark 3.2 Even though Theorem 3.2 demonstrates that the regularized Wirtinger flow endows with a linear convergence rate, it requires extra regularization added on the objective function, and the step size, i.e., $\eta \lesssim \frac{1}{s\kappa m}$ [10], lacks of aggressiveness. To exclude the regularization and achieve more aggressive step size, the papers [6, 7] have recently investigated regularization-free Wirtinger flow which yields a more aggressive step size, i.e., $\eta \lesssim s^{-1}$.

3.4.2 Regularization-Free Wirtinger Flow

Another line of studies has focused on the blind demixing model that is in the form of the bilinear model (3.13). In this section, we would formulate an optimization problem concerning the bilinear formulation of blind demixing and introduce efficient regularizer-free algorithms. The theoretical analysis on these algorithms will also be discussed.

A least-squares optimization problem under the scheme of the bilinear formulation of blind demixing is given by

$$\underset{\{\boldsymbol{h}_i\},\{\boldsymbol{x}_i\}}{\text{minimize}}\ f(\boldsymbol{h},\boldsymbol{x}) := \sum_{j=1}^{m}\Big|\sum_{i=1}^{s}\boldsymbol{b}_j^{\mathsf{H}}\boldsymbol{h}_i\boldsymbol{x}_i^{\mathsf{H}}\boldsymbol{a}_{ij} - y_j\Big|^2. \tag{3.28}$$

For simplification, the objective function in (3.28) is denoted as

$$f(\boldsymbol{z}) := f(\boldsymbol{h},\boldsymbol{x}),$$

where

$$\boldsymbol{z} = \left[\boldsymbol{z}_1^{\mathsf{H}} \cdots \boldsymbol{z}_s^{\mathsf{H}}\right]^{\mathsf{H}} \in \mathbb{C}^{2sK} \text{ with } \boldsymbol{z}_i = \left[\boldsymbol{h}_i^{\mathsf{H}}\ \boldsymbol{x}_i^{\mathsf{H}}\right]^{\mathsf{H}} \in \mathbb{C}^{2K}.$$

A line of literatures, e.g., [5–7, 10], have developed effective algorithms to solve problem (3.28). The blind demixing problem can be generally solved via two procedures [6, 10], i.e., Stage I: find an initial point that is in the neighborhood of the ground truth, which can be accomplished via spectral initialization; Stage II: optimize the initial estimate via an iterative algorithm, e.g., Wirtinger flow:

$$\begin{bmatrix}\boldsymbol{h}_i^{t+1}\\ \boldsymbol{x}_i^{t+1}\end{bmatrix} = \begin{bmatrix}\boldsymbol{h}_i^{t}\\ \boldsymbol{x}_i^{t}\end{bmatrix} - \eta\begin{bmatrix}\frac{1}{\|\boldsymbol{x}_i^t\|_2^2}\nabla_{\boldsymbol{h}_i} f(\boldsymbol{z}^t)\\ \frac{1}{\|\boldsymbol{h}_i^t\|_2^2}\nabla_{\boldsymbol{x}_i} f(\boldsymbol{z}^t)\end{bmatrix}, i = 1,\ldots,s, \tag{3.29}$$

where $\eta > 0$ is the step size, $\nabla_{\boldsymbol{h}_i} f(\boldsymbol{z})$ and $\nabla_{\boldsymbol{x}_i} f(\boldsymbol{z})$ represent the Wirtinger gradient of $f(\boldsymbol{z})$ with respect to $\boldsymbol{h}_i$ and $\boldsymbol{x}_i$, respectively.

The discrepancy between the estimate $\boldsymbol{z}$ and the ground truth $\boldsymbol{z}^\natural$ is defined as the distance function:

$$\mathrm{dist}(\boldsymbol{z},\boldsymbol{z}^\natural) = \left(\sum_{i=1}^{s}\mathrm{dist}^2(\boldsymbol{z}_i,\boldsymbol{z}_i^\natural)\right)^{1/2}, \tag{3.30}$$

Table 3.3 Conditions involved in Theorem 3.3 and corresponding references

Conditions	Reference
Incoherence between $\boldsymbol{a}_j$ and $\boldsymbol{x}_i$	(6b) in [6]
Incoherence between $\boldsymbol{b}_j$ and $\boldsymbol{h}_i$	(6c) in [6]
Robustness condition: $\|\mathscr{A}^*(\boldsymbol{e})\| \leq \gamma_e$	Sect. 6.5 in [10]

where

$$\mathrm{dist}^2(z_i, z_i^\natural) = \min_{\alpha_i \in \mathbb{C}} \left(\|\frac{1}{\overline{\alpha_i}}\boldsymbol{h}_i - \boldsymbol{h}_i^\natural\|_2^2 + \|\alpha_i \boldsymbol{x}_i - \boldsymbol{x}_i^\natural\|_2^2 \right) / d_i$$

for $i = 1, \ldots, s$. Here, $d_i = \|\boldsymbol{h}_i^\natural\|_2^2 + \|\boldsymbol{x}_i^\natural\|_2^2$ and each α_i is an alignment parameter. Without loss of generality, the ground-truth vectors are assumed to obey $\|\boldsymbol{h}_i^\natural\|_2 = \|\boldsymbol{x}_i^\natural\|_2$ for $i = 1, \ldots, s$. Recall the operator in (3.19) such that

$$\mathscr{A}_i^*(\boldsymbol{e}) = \sum_{j=1}^{m} e_j \boldsymbol{b}_j \boldsymbol{a}_{ij}^{\mathsf{H}}, \ i = 1, \ldots, s,$$

and the condition number κ in (3.25), then the theorem of Wirtinger flow with the spectral initialization for solving the blind demixing problem is presented in Theorem 3.3.

For simplicity, we summarize the conditions involved in the analysis of solving the blind demixing problem (3.28) via Wirtinger flow with spectral initialization in Table 3.3, and the detailed formulations of these parameters can be referred to the references presented in the table.

Theorem 3.3 *Assuming that the step size obeys* $\eta > 0$, $\eta \asymp s^{-1}$, *and the conditions in Table 3.3 are satisfied, the iterates (including the spectral initialization point) in Wirtinger flow satisfy*

$$\mathrm{dist}(z^t, z^\natural) \leq C_1 \left(1 - \frac{\eta}{16\kappa}\right)^t \left(\frac{1}{\log^2 m} - \frac{48\sqrt{s}\kappa^2}{\eta} \cdot \gamma_e\right) + \frac{48 C_1 \sqrt{s}\kappa^2}{\eta} \gamma_e,$$

for all $t \geq 0$, *with high probability if the number of measurements satisfies*

$$m \geq C(\mu^2 + \sigma^2) s^2 \kappa^4 K \log^8 m$$

for some constants $C_1 > 0$ *and adequately large constant* $C > 0$.

Proof Please refer to Sect. 8.3 for details.

Theorem 3.3 provides the justification for a more aggressive step size (i.e., $\eta \asymp s^{-1}$) even without regularization, compared to the step size (i.e., $\eta \lesssim \frac{1}{s\kappa m}$) given in [10] for regularized Wirtinger flow. In addition, the performance of the Wirtinger flow algorithm with spectral initialization is illustrated in Fig. 3.2a, b, which is endorsed

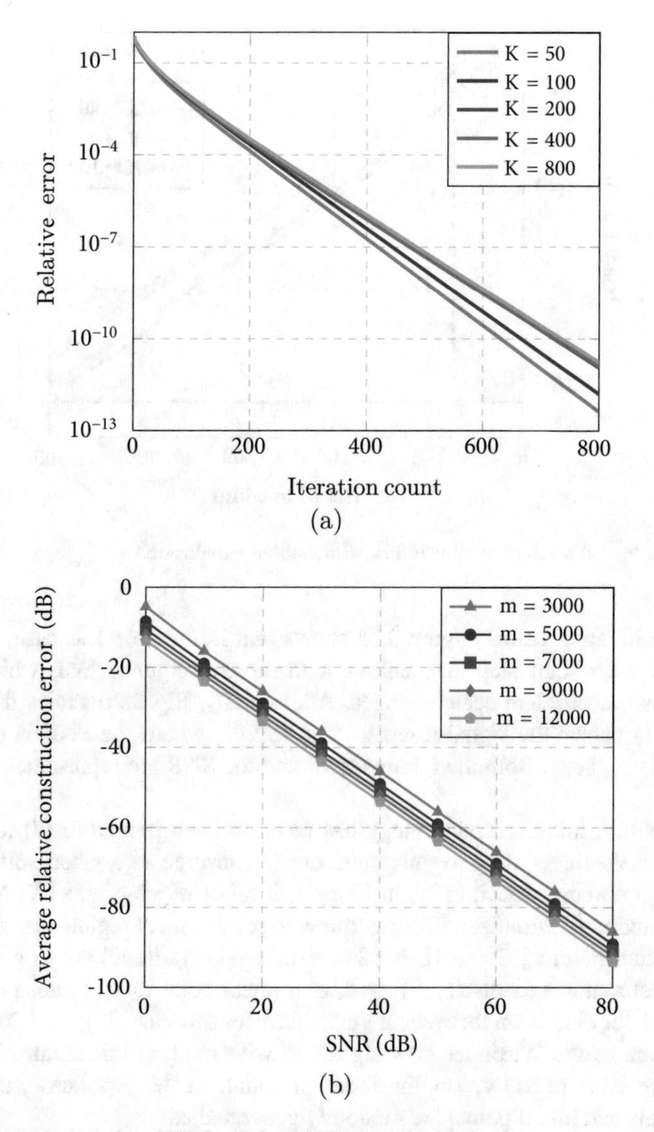

Fig. 3.2 Numerical results of Wirtinger flow with spectral initialization

by Theorem 3.3. To be specific, for each $K \in \{50, 100, 200, 400, 800\}$, $s = 10$, and $m = 50K$, the design vectors $\boldsymbol{a}_{ij}$'s and $\boldsymbol{b}_j$'s for each $1 \leq i \leq s, 1 \leq j \leq m$ are generated based on the instructions in Sect. 3.2.3. The underlying signals $\boldsymbol{h}_i^\natural, \boldsymbol{x}_i^\natural \in \mathbb{C}^K$, $1 \leq i \leq s$, are generated as random vectors with unit norm. With the chosen step size $\eta = 0.1$ in all settings, Fig. 3.2a shows the relative error, i.e.,

$$\frac{\sum_{i=1}^{s} \|\boldsymbol{h}_i^t \boldsymbol{x}_i^{t\mathsf{H}} - \boldsymbol{h}_i^\natural \boldsymbol{x}_i^{\natural\mathsf{H}}\|_F}{\sum_{i=1}^{s} \|\boldsymbol{h}_i^\natural \boldsymbol{x}_i^{\natural\mathsf{H}}\|_F}, \tag{3.31}$$

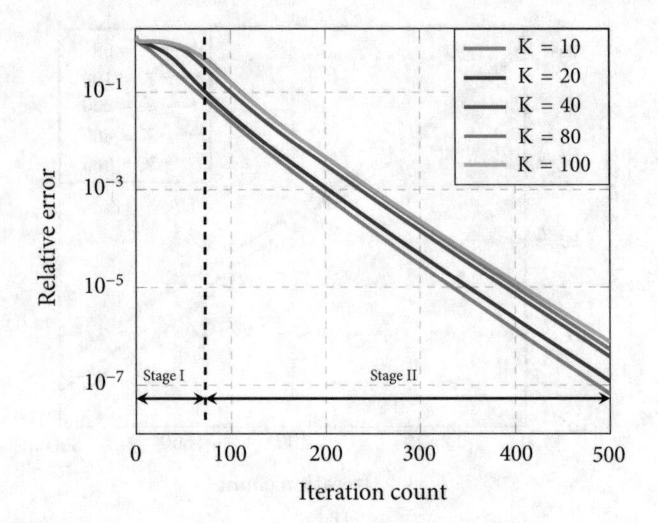

Fig. 3.3 Numerical result of Wirtinger flow with random initialization

versus the iteration count. Figure 3.2a shows that in the noiseless case, Wirtinger flow with a constant step size enjoys a linear convergence rate, which barely changes as the problem scale changes. Additionally, Fig. 3.2b shows the relative error (3.31) versus the signal-to-noise ratio (SNR), where the SNR is defined as SNR $:= \|\boldsymbol{y}\|_2/\|\boldsymbol{e}\|_2$. Both the relative error and the SNR are represented in the dB scale.

The random initialization strategy has recently been proven in [7] to be good enough for Wirtinger flow to guarantee linear converge rate when solving blind demixing problems. Specifically, in Stage I, it takes $\mathcal{O}(s\log(\max\{K, N\}))$ iterations for randomly initialized Wirtinger flow to reach a local region near the ground truth. Furthermore, in Stage II, it takes $\mathcal{O}(s\log(1/\varepsilon))$ iterations to attain an ε-accurate estimator, i.e., $\text{dist}(\boldsymbol{z}, \boldsymbol{z}^{\natural}) \leq \epsilon$, at a linear convergence rate. Please refer to Sect. 8.4 for details on theoretical guarantees for this case. Figure 3.3 shows the performance of the Wirtinger flow algorithm with random initialization, showing the relative error (3.31) versus the iteration count. In the simulation, the ground truth signals and initial points are randomly generated as

$$\boldsymbol{h}_i^{\natural} \sim \mathcal{CN}(\boldsymbol{0}, K^{-1}\boldsymbol{I}_K),\ \boldsymbol{x}_i^{\natural} \sim \mathcal{CN}(\boldsymbol{0}, N^{-1}\boldsymbol{I}_N), \tag{3.32}$$

$$\boldsymbol{h}_i^{0} \sim \mathcal{CN}(\boldsymbol{0}, K^{-1}\boldsymbol{I}_K),\ \boldsymbol{x}_i^{0} \sim \mathcal{CN}(\boldsymbol{0}, N^{-1}\boldsymbol{I}_N), \tag{3.33}$$

for $i = 1, \ldots, s$. In all simulations, we set $K = N$ for each $K \in \{10, 20, 40, 80, 100\}$, $s = 10$, and $m = 50K$, and with the chosen step size $\eta = 0.1$.

The above nonconvex algorithm has a low iteration cost, and the overall computational complexity can be further decreased via reducing the iteration complexity, i.e., accelerating the convergence rate. This motivates to develop the Riemannian optimization algorithm which will be introduced in the next section.

3.4.3 Riemannian Optimization Algorithm

The paper [8] developed a Riemannian trust-region algorithm on the complex product manifolds to solve the blind demixing problem, which enjoys a fast convergence rate. Prior to introducing this algorithm for solving the blind demixing problem, we start with some basic concepts of Riemannian manifold optimization, and readers can refer to Sect. 8.5 in the book [1] for more details.

3.4.3.1 An Example on Riemannian Optimization

In order to optimize a smooth function on a manifold, several geometric concepts in terms of manifolds are required. To be specific, tangent vectors on manifolds generalize the notion of a direction, and an inner product of tangent vectors generalizes a notion of length that applies to these tangent vectors. A *Riemannian manifold*, generally denoted as $\mathcal{M}$, is the manifold of which tangent spaces $T_{\boldsymbol{x}}\mathcal{M}$ are endowed with a smoothly varying inner product. The smoothly varying inner product is called the *Riemannian metric*, generally denoted as

$$g_{\boldsymbol{x}}(\boldsymbol{\eta}_{\boldsymbol{x}}, \boldsymbol{\zeta}_{\boldsymbol{x}}),$$

where $\boldsymbol{x} \in \mathcal{M}$ and $\boldsymbol{\eta}_{\boldsymbol{x}}, \boldsymbol{\zeta}_{\boldsymbol{x}} \in T_{\boldsymbol{x}}\mathcal{M}$. Some examples of Riemannian manifold can be enumerated as: sphere, orthogonal Stiefel manifold, Grassmann manifold, rotation group, positive definite matrices, fixed-rank matrices, etc.

Consider minimizing a smooth function on the sphere $\mathbb{S}^{n-1} = \{\boldsymbol{x} \in \mathbb{R}^n : \boldsymbol{x}^\top \boldsymbol{x} = 1\}$:

$$\underset{\boldsymbol{x}\in\mathbb{R}^n}{\text{minimize}}\ f(\boldsymbol{x}) = -\boldsymbol{x}^\top \boldsymbol{A}\boldsymbol{x} \quad \text{subject to} \quad \boldsymbol{x}^\top \boldsymbol{x} = 1, \tag{3.34}$$

where $\boldsymbol{A}$ is a symmetric matrix. As illustrated in Fig. 3.4, the Riemannian optimization procedure on the sphere can be separated into three steps:

1. Compute the Euclidean gradient in $\mathbb{R}^n$:

$$\nabla f(\boldsymbol{x}) = -2\boldsymbol{A}\boldsymbol{x}. \tag{3.35}$$

2. Compute the Riemannian gradient on the sphere $\mathbb{S}^{n-1}$ via projecting $\nabla f(\boldsymbol{x})$ to the tangent space $T_{\boldsymbol{x}}\mathcal{M}$:

$$\operatorname{grad} f(\boldsymbol{x}) = \operatorname{Proj}_{\boldsymbol{x}} \nabla f(\boldsymbol{x}) = (\boldsymbol{I} - \boldsymbol{x}\boldsymbol{x}^\top)\nabla f(\boldsymbol{x}). \tag{3.36}$$

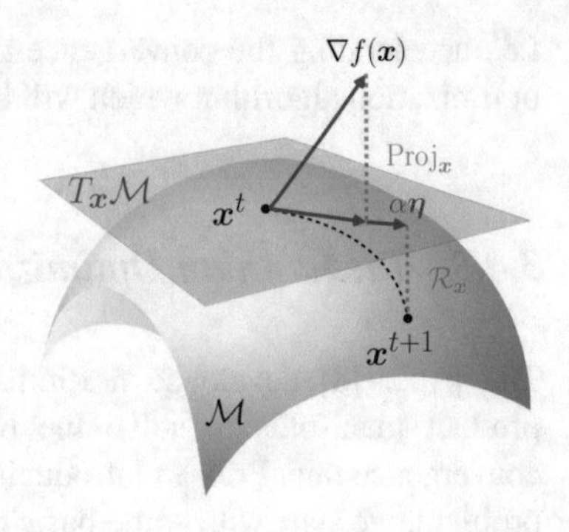

Fig. 3.4 Schematic viewpoint of Riemannian optimization on the Riemannian manifold

3. Move along the descent direction $\boldsymbol{\eta} = \operatorname{grad} f(\boldsymbol{x})$ and retract the directional vector $\alpha\boldsymbol{\eta}$ to the sphere, where $\alpha > 0$ is the step size. The retraction operator on $\mathbb{S}^{n-1}$ is given by

$$\mathscr{R}_{\boldsymbol{x}}(\alpha\boldsymbol{\eta}) = \mathrm{qf}(\boldsymbol{x} + \alpha\boldsymbol{\eta}), \tag{3.37}$$

where $\mathrm{qf}(\cdot)$ denotes the mapping that maps a matrix to the Q factor of its QR decomposition.

Furthermore, for the ease of implementing the optimization scheme on manifolds, a powerful Matlab toolbox, namely Manopt [2], has been developed, which contains a larger library of manifolds (e.g., sphere, orthogonal Stiefel manifold, Grassmann manifold, rotation group, positive definite matrices, fixed-rank matrices, etc.) and various Riemannian optimization algorithms (e.g., steepest descent, conjugate gradient, stochastic gradient descent, trust-regions algorithm, etc.).

3.4.3.2 Riemannian Optimization on Product Manifolds for Blind Demixing

Due to the multiple rank-one matrices in the blind demixing problem (3.20), problem (3.20) can be reformulated as minimizing a smooth function on the product of multiple fixed-rank matrices. The product of multiple fixed-rank matrices is a product manifold and is also a Riemannian manifold [1]. The example mentioned above paves the way for dealing with more complicated Riemannian optimization algorithms on product manifolds for solving the blind demixing problem.

Firstly, a linear map is developed to handle complex asymmetric matrices. The linear map facilitates to convert the optimization variables to a Hermitian positive semidefinite matrix. Define a linear map

$$\mathscr{J}_i : \mathbb{S}_+^{(N+K)} \to \mathbb{C}^m$$

with respect to a Hermitian positive semidefinite (PSD) matrix $\boldsymbol{Y}_i$ that obeys

$$[\mathscr{J}_i(\boldsymbol{Y}_i)]_i = \langle \boldsymbol{J}_{ij}, \boldsymbol{Y}_i \rangle \tag{3.38}$$

with $\boldsymbol{Y}_i \in \mathbb{S}_+^{(N+K)}$ and $\boldsymbol{J}_{ij}$ as

$$\boldsymbol{J}_{ij} = \begin{bmatrix} \boldsymbol{0}_{N \times N} & \boldsymbol{A}_{ij} \\ \boldsymbol{0}_{K \times N} & \boldsymbol{0}_{K \times K} \end{bmatrix} \in \mathbb{C}^{(N+K)\times(N+K)}, \tag{3.39}$$

where $\boldsymbol{A}_{ij} = \boldsymbol{a}_{ij}\bar{\boldsymbol{b}}_j^{\mathsf{H}}$. Note that based on (3.38), we have

$$[\mathscr{J}_i(\boldsymbol{M}_i)]_i = \langle \boldsymbol{J}_{ij}, \boldsymbol{M}_i \rangle = \langle \boldsymbol{A}_{ij}, \boldsymbol{x}_i \bar{\boldsymbol{h}}_i^{\mathsf{H}} \rangle, \tag{3.40}$$

where $\boldsymbol{M}_i = \boldsymbol{w}_i \boldsymbol{w}_i^{\mathsf{H}}$ with $\boldsymbol{w}_i = [\boldsymbol{x}_i^{\mathsf{H}}\ \bar{\boldsymbol{h}}_i^{\mathsf{H}}]^{\mathsf{H}} \in \mathbb{C}^{N+K}$. Based on the matrix factorization, a manifold optimization problem with respect to Hermitian positive semidefinite (PSD) matrices can be established as:

$$\underset{\boldsymbol{v}=\{\boldsymbol{w}_k\}_{k=1}^s}{\text{minimize}}\ f(\boldsymbol{v}) := \left\| \sum\nolimits_{k=1}^{s} \mathscr{J}_k\big(\boldsymbol{w}_k \boldsymbol{w}_k^{\mathsf{H}}\big) - \boldsymbol{y} \right\|^2, \tag{3.41}$$

where $\boldsymbol{v} \in \mathscr{M}^s$ with $\boldsymbol{w}_k \in \mathscr{M} := \mathbb{C}_*^{N+K}$ for $k = 1, \ldots, s$, where the space $\mathbb{C}_*^n$ is the complex Euclidean space $\mathbb{C}^n$ without the origin. According to (3.40), the data signal estimation $\hat{\boldsymbol{x}}_k$ can be represented by the first N rows of the estimation $\hat{\boldsymbol{w}}_k$.

Since the quotient manifold is abstract, the matrix representations in the computational space $\mathscr{M}^s$ of the geometric concepts in the quotient space are required. In particular, to develop the Riemannian optimization algorithm over the product manifolds, various geometric concepts need to be derived, such as the notion of length (i.e., Riemannian metric $g_{\boldsymbol{w}_k}$), set of directional derivatives (i.e., horizontal space $\mathscr{H}_{\boldsymbol{w}_k}$), and motion along geodesics (i.e., retraction $\mathscr{R}_{\boldsymbol{w}_k}$) [1]. The concrete optimization-related ingredients are shown in Table 3.4. Based on these ingredients, we develop a Riemannian algorithm to solve the blind demixing problem (3.41).

Based on the geometry of the product manifolds, the Riemannian optimization algorithm operated on the product manifolds $\mathscr{M}^s$ can be elementwisely developed on the individual manifold $\mathscr{M}$. To be specific, for each $k = 1, 2, \ldots, s$, the descent direction $\boldsymbol{\eta}$ is detected on the horizontal space $\mathscr{H}_{\boldsymbol{w}_k}\mathscr{M}$ parallelly, and $\boldsymbol{\eta}$ is parallelly retracted on the individual manifold $\mathscr{M}$ via the retraction mapping $\mathscr{R}_{\boldsymbol{w}_k}$. In addition, Fig. 3.5 shows the schematic viewpoint of Algorithm 3.2.

Riemannian Gradient Descent with Spectral Initialization In the Riemannian gradient descent algorithm, i.e., Algorithm 3.3, the search direction is given by

$$\boldsymbol{\eta} = -\text{grad}_{\boldsymbol{w}_k^{[t]}} f / g_{\boldsymbol{w}_k^{[t]}}\big(\boldsymbol{w}_k^{[t]}, \boldsymbol{w}_k^{[t]}\big),$$

Table 3.4 Elementwise optimization-related ingredients for Problem (3.41)

	$\mathrm{Minimize}_{\boldsymbol{w}_k \in \mathcal{M}} \left\| \sum_{k=1}^{s} \mathcal{J}_k(\boldsymbol{w}_k \boldsymbol{w}_k^{\mathsf{H}}) - \boldsymbol{y} \right\|^2$
Computational space: $\mathcal{M}$	$\mathbb{C}_*^{N+K}$
Quotient space: $\mathcal{M}/\sim$	$\mathbb{C}_*^{N+K}/\mathrm{SU}(1)$
Riemannian metric: $g_{\boldsymbol{w}_k}$	$g_{\boldsymbol{w}_k}(\boldsymbol{\zeta}_{\boldsymbol{w}_k}, \boldsymbol{\eta}_{\boldsymbol{w}_k}) = \mathrm{Tr}(\boldsymbol{\zeta}_{\boldsymbol{w}}{}^{\mathsf{H}}{}_k \boldsymbol{\eta}_{\boldsymbol{w}_k} + \boldsymbol{\eta}_{\boldsymbol{w}}{}^{\mathsf{H}}{}_k \boldsymbol{\zeta}_{\boldsymbol{w}_k})$
Horizontal space: $\mathcal{H}_{\boldsymbol{w}_k}\mathcal{M}$	$\boldsymbol{\eta}_{\boldsymbol{w}_k} \in \mathbb{C}^{N+K} : \boldsymbol{\eta}_{\boldsymbol{w}_k}^{\mathsf{H}} \boldsymbol{w}_k = \boldsymbol{w}_k^{\mathsf{H}} \boldsymbol{\eta}_{\boldsymbol{w}_k}$
Horizontal space projection	$\Pi_{\mathcal{H}_{\boldsymbol{w}_k}\mathcal{M}}(\boldsymbol{\eta}_{\boldsymbol{w}_k}) = \boldsymbol{\eta}_{\boldsymbol{w}_k} - a\boldsymbol{w}_k$, $a = (\boldsymbol{w}_k^{\mathsf{H}} \boldsymbol{\eta}_{\boldsymbol{w}_k} - \boldsymbol{\eta}_{\boldsymbol{w}_k}^{\mathsf{H}} \boldsymbol{w}_k)/2\boldsymbol{w}^{\mathsf{H}}\boldsymbol{w}$
Riemannian gradient: $\mathrm{grad}_{\boldsymbol{w}_k} f$	$\mathrm{grad}_{\boldsymbol{w}} f = \Pi_{\mathcal{H}_{\boldsymbol{w}_k}\mathcal{M}}\left(\frac{1}{2}\nabla_{\boldsymbol{w}_k} f(\boldsymbol{v})\right)$
Riemannian Hessian: $\mathrm{Hess}_{\boldsymbol{w}_k} f[\boldsymbol{\eta}_{\boldsymbol{w}_k}]$	$\mathrm{Hess}_{\boldsymbol{w}_k} f[\boldsymbol{\eta}_{\boldsymbol{w}_k}] = \Pi_{\mathcal{H}_{\boldsymbol{w}_k}\mathcal{M}}\left(\frac{1}{2}\nabla^2_{\boldsymbol{w}_k} f(\boldsymbol{v})[\boldsymbol{\eta}_{\boldsymbol{w}_k}]\right)$
Retraction: $\mathcal{R}_{\boldsymbol{w}_k} : T_{\boldsymbol{w}_k}\mathcal{M} \to \mathcal{M}$	$\mathcal{R}_{\boldsymbol{w}_k}(\boldsymbol{\eta}_{\boldsymbol{w}_k}) = \boldsymbol{w}_k + \boldsymbol{\eta}_{\boldsymbol{w}_k}$

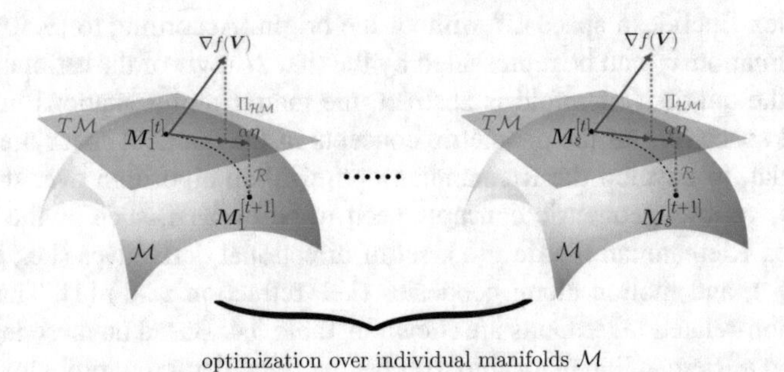

Fig. 3.5 Schematic viewpoint of Riemannian optimization on the product manifolds

where $g_{\boldsymbol{w}_k^{[t]}}$ is the Riemannian metric and

$$\mathrm{grad}_{\boldsymbol{w}_k^{[t]}} f \in \mathcal{H}_{\boldsymbol{w}_k}\mathcal{M}$$

is the Riemannian gradient. Therefore, the sequence of the iterates is given by

$$\boldsymbol{w}_k^{[t+1]} = \mathcal{R}_{\boldsymbol{w}_k^{[t]}}(\alpha_t \boldsymbol{\eta}),$$

Algorithm 3.2: Riemannian optimization on product manifolds

Given: Riemannian manifold $\mathcal{M}^s$ with Riemannian metric $g_{\boldsymbol{v}}$, retraction mapping $\mathcal{R}_{\boldsymbol{v}} = \{\mathcal{R}_{\boldsymbol{w}_k}\}_{k=1}^s$, objective function f and the step size α.
Output: $\boldsymbol{v} = \{\boldsymbol{w}_k\}_{k=1}^s$
1: Initialize: initial point $\boldsymbol{v}^{[0]} = \{\boldsymbol{w}_k^{[0]}\}_{k=1}^s$, $t = 0$
2: while not converged do
3: for all $k = 1, \cdots, s$ do in parallel
4: Compute a descent direction $\boldsymbol{\eta}$. (e.g., via implementing trust-region method)
5: Update $\boldsymbol{w}_k^{[t+1]} = \mathcal{R}_{\boldsymbol{w}_k^{[t]}}(\alpha\boldsymbol{\eta})$
6: $t = t + 1$.
7: end for
8: end while

where the step size $\alpha_t > 0$ and

$$\mathcal{R}_{\boldsymbol{w}_k}(\boldsymbol{\xi}) = \boldsymbol{w}_k + \boldsymbol{\xi}, \tag{3.42}$$

with $\boldsymbol{\xi} \in \mathcal{H}_{\boldsymbol{w}_k}\mathcal{M}$. Here, the retraction map

$$\mathcal{R}_{\boldsymbol{w}_k} : \mathcal{H}_{\boldsymbol{w}_k}\mathcal{M} \to \mathcal{M}$$

is an approximation of the exponential map that characterizes the motion of "moving along geodesics on the Riemannian manifold." More details on computing the retraction are available in [1, Section 4.1.2]. The statistical analysis of the Riemannian gradient descent algorithm will be provided in the sequel, which demonstrates the linear rate of the proposed algorithm for converging to the ground truth signals.

Theorem 3.4 *Suppose the rows of the encoding matrices, i.e., $\boldsymbol{c}_{ij}$'s, follow the i.i.d. complex Gaussian distribution, i.e.,*

$$\boldsymbol{c}_{ij} \sim \mathcal{N}\left(0, \frac{1}{2}\boldsymbol{I}_N\right) + i\mathcal{N}\left(0, \frac{1}{2}\boldsymbol{I}_N\right)$$

and the step size obeys $\alpha_t > 0$ and $\alpha_t \equiv \alpha \asymp s^{-1}$, then the iterates (including the spectral initialization point) in Algorithm 3.3 satisfy

$$\mathrm{dist}(\boldsymbol{v}^t, \boldsymbol{v}^\natural) \leq C_1\left(1 - \frac{\alpha}{16\kappa}\right)^t \frac{1}{\log^2 L} \tag{3.43}$$

for all $t \geq 0$ and some constant $C_1 > 0$, with probability at least $1 - c_1 L^{-\gamma} - c_1 L e^{-c_2 K}$ if the number of measurements

$$L \geq C\mu^2 s^2 \kappa^4 \max\{K, N\} \log^8 L$$

for some constants $\gamma, c_1, c_2 > 0$ and sufficiently large constant $C > 0$.

Algorithm 3.3: Riemannian gradient descent with spectral initialization

Given: Riemannian manifold $\mathcal{M}^s$ with optimization-related ingredients, objective function f, $\{\boldsymbol{c}_{ij}\}$, $\{\boldsymbol{b}_j\}$, $\{y_j\}$ and the stepsize α.
Output: $\boldsymbol{v} = \{\boldsymbol{w}_k\}_{k=1}^s$
1: Spectral Initialization:
2: for all $i = 1, \cdots, s$ do in parallel
3: Let $\sigma_1(\boldsymbol{N}_i)$, $\check{\boldsymbol{h}}_i^0$ and $\check{\boldsymbol{x}}_i^0$ be the leading singular value, left singular vector and right singular vector of matrix $\boldsymbol{N}_i := \sum_{j=1}^m y_j \boldsymbol{b}_j \boldsymbol{c}_{ij}^{\mathsf{H}}$, respectively.
4: Set $\boldsymbol{w}_i^{[0]} = \begin{bmatrix} \boldsymbol{x}_i^0 \\ \boldsymbol{h}_i^0 \end{bmatrix}$ where $\boldsymbol{x}_i^0 = \sqrt{\sigma_1(\boldsymbol{N}_i)}\check{\boldsymbol{x}}_i^0$ and $\boldsymbol{h}_i^0 = \sqrt{\sigma_1(\boldsymbol{N}_i)}\check{\boldsymbol{h}}_i^0$.
5: end for
6: for all $t = 1, \cdots, T$
7: for all $i = 1, \cdots, s$ do in parallel
8: $\boldsymbol{\eta} = -\frac{1}{g_{\boldsymbol{w}_k^{[t]}}(\boldsymbol{w}_k^{[t]}, \boldsymbol{w}_k^{[t]})} \operatorname{grad}_{\boldsymbol{w}_k^{[t]}} f$
9: Update $\boldsymbol{w}_k^{[t+1]} = \mathscr{R}_{\boldsymbol{w}_k^{[t]}}(\alpha_t \boldsymbol{\eta})$
10: end for
11: end for

Proof Please refer to Sect. 8.6 for details.

Theorem 3.4 demonstrates that the number of measurements $\mathcal{O}(s^2\kappa^4 \max\{K, N\} \log^8 L)$ are sufficient for the Riemannian gradient descent algorithm (with spectral initialization), i.e., Algorithm 3.3, to linearly converge to the ground truth signals.

Riemannian Trust-Region Algorithm A scalable algorithm that enjoys superlinear convergence rate, i.e., the Riemannian trust-region algorithm, can be developed on the product manifolds to detect the descent direction $\boldsymbol{\eta}$ [1, Section 7]. In order to parallelly search the descent direction on the horizontal space $\mathscr{H}_{\boldsymbol{v}}\mathcal{M}^s$, the method of searching the direction $\boldsymbol{\eta}_{\boldsymbol{w}_k}$ on the horizontal space $\mathscr{H}_{\boldsymbol{w}_k}\mathcal{M}$ is developed. At each iteration, define the point on the manifold as $\boldsymbol{w}_k \in \mathcal{M}$, a trust-region subproblem is described as follows [1]:

$$\begin{aligned} \underset{\boldsymbol{\eta}_{\boldsymbol{w}_k}}{\text{minimize}} \quad & m(\boldsymbol{\eta}_{\boldsymbol{w}_k}) \\ \text{subject to} \quad & g_{\boldsymbol{w}_k}(\boldsymbol{\eta}_{\boldsymbol{w}_k}, \boldsymbol{\eta}_{\boldsymbol{w}_k}) \leq \delta^2, \end{aligned} \tag{3.44}$$

where $\boldsymbol{\eta}_{\boldsymbol{w}_k} \in \mathscr{H}_{\boldsymbol{w}_k}\mathcal{M}$, δ is the trust-region radius, and the objective function is represented as

$$m(\boldsymbol{\eta}_{\boldsymbol{w}_k}) = g_{\boldsymbol{w}_k}(\boldsymbol{\eta}_{\boldsymbol{w}_k}, \operatorname{grad}_{\boldsymbol{w}_k} f) + \frac{1}{2} g_{\boldsymbol{w}_k}\left(\boldsymbol{\eta}_{\boldsymbol{w}_k}, \operatorname{Hess}_{\boldsymbol{w}_k} f\left[\boldsymbol{\eta}_{\boldsymbol{w}_k}\right]\right), \tag{3.45}$$

and $\operatorname{Hess}_{\boldsymbol{w}_k} f\left[\boldsymbol{\eta}_{\boldsymbol{w}_k}\right]$ and $\operatorname{grad}_{\boldsymbol{w}_k} f$ are the matrix representations of Riemannian Hessian and Riemannian gradient in the quotient space, respectively. In addition, the iterate being updated or maintained depends on whether the decrease of the function $m(\eta_{w_k})$ is satisfied or not [1, Section 7]. If the decrease is sufficient, the iterate is updated as

$$\mathscr{R}_{w_k}(\eta_{w_k}) = w_k + \eta_{w_k}. \tag{3.46}$$

Under the above framework, the Riemannian trust-region algorithm is parallelly developed on individual manifolds to solve problem (3.41).

3.4.4 Simulation Results

In this section, we compare three algorithms for estimating the blind demixing model: nuclear norm minimization (NNM) in Sect. 3.3.1, regularized Wirtinger flow (RGD) in Sect. 3.4.1, and Riemannian trust-region algorithm (RTR) in Sect. 3.4.3.

The ground-truth vectors, i.e., $x_k \in \mathbb{C}^N$ and $h_k \in \mathbb{C}^K$ for $k = 1, \ldots, s$, are generated as standard complex Gaussian vectors whose entries are drawn i.i.d. from the standard normal distribution. In addition, the relative construction error with respect to the rank-one matrices, i.e., $X_i = h_i x_i^{\mathsf{H}}$, is adopted to evaluate the performance of the algorithms, given as

$$\mathrm{err}(X) = \frac{\sqrt{\sum_{k=1}^{s} \|X_k - \hat{X}_k\|_F^2}}{\sqrt{\sum_{k=1}^{s} \|\hat{X}_k\|_F^2}}, \tag{3.47}$$

where $\{X_k\}$ are estimated matrices and $\{\hat{X}_k\}$ are ground truth matrices. The initialization strategy, i.e., Algorithm 3.1, is adopted for all the nonconvex optimization algorithms, i.e., RGD and RTR. The RTR algorithm stops when the norm of Riemannian gradient is less than 10^{-8} or the number of iterations exceeds 500. The stopping criteria of RGD is adopted from the paper [10].

In the noiseless scenario, two nonconvex algorithms are compared under the setting of $N = K = 50$, $L = 1250$, and $s = 5$. The convergence rates of nonconvex algorithms are illustrated in Fig. 3.6. In the noisy scenario, assume the additive noise term in (3.17) obeys

$$e = \sigma \cdot \|y\| \cdot \frac{\omega}{\|\omega\|}, \tag{3.48}$$

where $\omega \in \mathbb{C}^L$ denotes a standard complex Gaussian vector. Three algorithms with respect to different signal-to-noise ratios (SNR) σ are compared under the setting of $L = 1500$, $N = K = 50$, and $s = 2$. In each circumstance, ten independent trails are simulated. Figure 3.7 shows the average relative construction error in dB against the signal-to-noise ratio (SNR). It concludes that the average relative construction

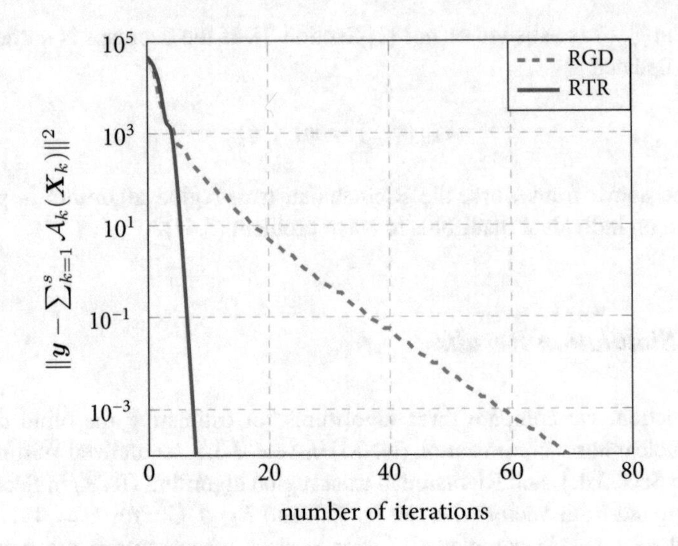

Fig. 3.6 Convergence rate of nonconvex algorithms

error decreases as SNR increases, which demonstrates that RTR is robust to the noise.

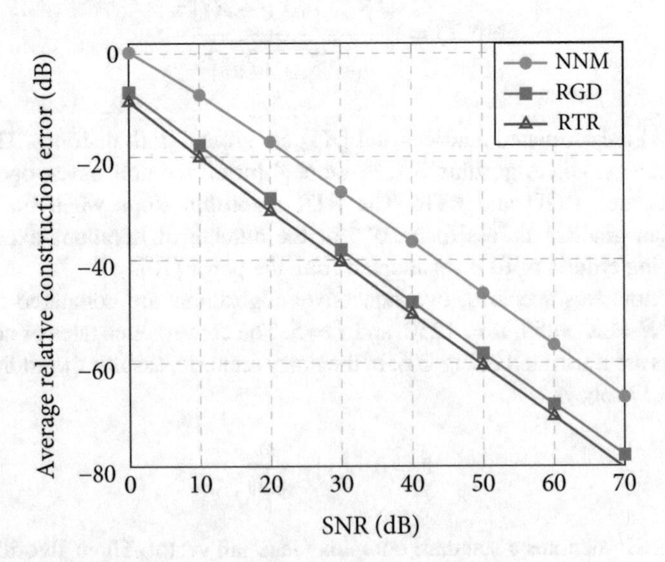

Fig. 3.7 Average relative construction error versus SNR (dB)

Table 3.5 Summary of approaches for solving blind demixing problem

	Convex relaxation	Nonconvex approach			
Algorithm	NNM Sect. 3.3.1	RGD Sect. 3.4.1	RTR Sect. 3.4.3	WF Sect. 3.4.2	WF Sect. 3.4.2
Regularizer	×	√	×	×	×
Initialization	×	Spectral	Spectral	Spectral	Random
Condition number	1	κ	κ	κ	κ
Sample complexity	$m \geq Cs^2\mu^2$ $K \log^2 m$	$m \geq Cs^2\mu^2$ $\kappa^4 K \log^2 m$	$m \geq Cs^2\mu^2$ $\kappa^4 K \log^8 m$	$m \geq Cs^2\mu^2$ $\kappa^4 K \log^8 m$	$m \geq Cs^2\mu^2$ $\kappa^4 K \log^8 m$
Computational complexity	–	$\mathcal{O}\left(sm \log \frac{1}{\epsilon}\right)$	Not provided	$\mathcal{O}\left(s \log \frac{1}{\epsilon}\right)$	$\mathcal{O}\left(s \log K + s \log \frac{1}{\epsilon}\right)$

3.5 Summary

This chapter introduced a blind demixing model that facilitates to jointly decode data and estimate the channel state in IoT networks. The low-overhead communications can be achieved via the blind demixing model since it excludes the channel estimation sequence in the metadata. The convex relaxation method is introduced to solve the blind demixing problem based on its low-rank property. To further reduce the computational complexity, first-order algorithms, e.g., Wirtinger flow and regularized Wirtinger flow, have been developed. In addition, a Riemannian trust-region algorithm that enjoys faster convergence than the first-order algorithm has also been presented. The summary of both convex and nonconvex approaches for solving the blind demixing problem in the noiseless scenario is provided in Table 3.5.[1] State-of-the-art theoretical analysis is developed under the assumption of the Gaussian encoding matrices. It is intriguing to explore more general types of encoding matrices, e.g., sub-Gaussian matrices, in future works.

References

1. Absil, P.A., Mahony, R., Sepulchre, R.: Optimization Algorithms on Matrix Manifolds. Princeton University Press, Princeton (2009)
2. Boumal, N., Mishra, B., Absil, P.A., Sepulchre, R.: Manopt, a Matlab toolbox for optimization on manifolds. J. Mach. Learn. Res. **15**, 1455–1459 (2014). http://www.manopt.org
3. Candes, E.J., Romberg, J.K., Tao, T.: Stable signal recovery from incomplete and inaccurate measurements. Commun. Pure Appl. Math. **59**(8), 1207–1223 (2006)
4. Candes, E.J., Strohmer, T., Voroninski, V.: PhaseLift: exact and stable signal recovery from magnitude measurements via convex programming. Commun. Pure Appl. Math. **66**(8), 1241–1274 (2013)
5. Chi, Y., Lu, Y.M., Chen, Y.: Nonconvex optimization meets low-rank matrix factorization: an overview. arXiv preprint. arXiv:1809.09573 (2018)
6. Dong, J., Shi, Y.: Nonconvex demixing from bilinear measurements. IEEE Trans. Signal Process. **66**(19), 5152–5166 (2018)
7. Dong, J., Shi, Y.: Blind demixing via Wirtinger flow with random initialization. In: Proceedings of the International Conference on Artificial Intelligence and Statistics (AISTATS), vol. 89, pp. 362–370 (2019)
8. Dong, J., Yang, K., Shi, Y.: Blind demixing for low-latency communication. IEEE Trans. Wireless Commun. **18**(2), 897–911 (2019)
9. Ling, S., Strohmer, T.: Blind deconvolution meets blind demixing: algorithms and performance bounds. IEEE Trans. Inf. Theory **63**(7), 4497–4520 (2017)
10. Ling, S., Strohmer, T.: Regularized gradient descent: a nonconvex recipe for fast joint blind deconvolution and demixing. Inf. Inference: J. IMA **8**(1), 1–49 (2018)
11. Parvez, I., Rahmati, A., Guvenc, I., Sarwat, A.I., Dai, H.: A survey on low latency towards 5G: RAN, core network and caching solutions. arXiv preprint. arXiv:1708.02562 (2017)

[1] In Table 3.5, the parameters are consistent with the blind demixing model (3.13). The incoherence parameter μ is mentioned in Table 3.1. We assume that $K = N$ and there is some sufficient large constant $C > 0$.

Chapter 4
Sparse Blind Demixing

Abstract This chapter extends the models presented in Chaps. 2 and 3 to the scenario involving device activity detection. The new setting induces a sparse blind demixing model for developing methods for joint device activity detection, data decoding, and channel estimation in IoT networks. The signal model is first presented, in the scenario with either a single-antenna or multi-antenna BS. A convex relaxation approach is first introduced as a basic method to solve the nonconvex estimation problem. We further present a difference-of-convex-functions (DC) approach which turns out to be a powerful tool to solve the resulting sparse and low-rank optimization problem with matrix lifting. Furthermore, a smooth Riemannian optimization algorithm operating on the product manifold is introduced for solving the sparse blind demixing problem directly.

4.1 Joint Device Activity Detection, Data Decoding, and Channel Estimation

In Chap. 2, a sparse linear model has been developed for grant-free random access to jointly detect device activity and estimate channel state information. Under this scheme, pilot sequences are needed for activity detection, which lead to excess overhead for short packet communications. To avoid the transmission of pilot sequences, more powerful signal processing techniques are needed for data detection. Assuming the active device set is known, the blind demixing model has been introduced in Chap. 3 to achieve pilot-free communications in the massive IoT network via joint data decoding and channel estimation. To further account for the sporadic activity pattern in massive IoT networks, a *sparse blind demixing model* was proposed in [4, 6] to reduce the overhead during the transmission via joint device activity detection, data decoding, and channel estimation in a unified way.

Y. Shi et al., *Low-overhead Communications in IoT Networks*,
https://doi.org/10.1007/978-981-15-3870-4_4

Considering an IoT network containing one BS equipped with a single antenna, where only part (denoted as the set $\mathscr{S}$) of the devices are active, the sparse blind demixing model represented in the Fourier domain is given as

$$y_j = \sum\nolimits_{k\in\mathscr{S}} \boldsymbol{b}_j^{\mathsf{H}}\boldsymbol{h}_k\boldsymbol{x}_k^{\mathsf{H}}\boldsymbol{a}_{kj},\ 1 \le j \le m, \tag{4.1}$$

where

$$\boldsymbol{y} = [y_1, \dots, y_m]^{\top} \in \mathbb{C}^m$$

is the received signal at the BS represented in the Fourier domain, $\{\boldsymbol{b}_j\}$ are design vectors that indicate the Fourier transform operation, $\{\boldsymbol{a}_{kj}\}$ are design vectors that indicate the encoding procedure, and $\{\boldsymbol{h}_k\}$, $\{\boldsymbol{x}_k\}$ are channel signals and data signals, respectively. By detecting the active set $\mathscr{S}$ and the vectors $\{\boldsymbol{h}_k\}$, $\{\boldsymbol{x}_k\}$ for $k \in \mathscr{S}$ from the observation $\boldsymbol{y}$, device activity detection, data decoding, and channel estimation can be simultaneously achieved. This is a highly challenging problem.

In the sequel, we first introduce the problem formulation of the sparse blind demixing model. Various approaches for solving the corresponding nonconvex estimation problem are then introduced: (1) a convex relaxation approach based on the minimization of nuclear norms and ℓ_1/ℓ_2-norms, (2) a difference-of-convex (DC) function approach based on the minimization of DC objective functions. Along the discussion, we also identify theoretical analysis for the sparse blind demixing model as future research directions.

4.2 Problem Formulation

In this section, we present problem formulation for joint activity detection, data decoding, and channel estimation for both scenarios of single-antenna and multi-antenna BSs. Considering an IoT network consisting of one BS and s single-antenna devices with sporadic traffic, in each coherence block, only an unknown subset of devices are active, defined as $\mathscr{S} \subseteq \{1, 2, \dots, s\}$.

4.2.1 Single-Antenna Scenario

In the single-antenna BS scenario, the problem formulation of the sparse blind demixing model can be derived from the blind demixing model mentioned in Sect. 3.2 with an additional consideration of the sparse activity pattern. The data signal transmitted by the k-th user is denoted as $\boldsymbol{x}_k^{\natural} \in \mathbb{C}^N$. Assume that an encoding matrix over the m time slots is assigned to each device k. Over m time slots, the received signals at the BS in the frequency domain are presented as [3, 9]

$$y_j = \sum_{k\in\mathscr{S}} \boldsymbol{b}_j^{\mathsf{H}} \boldsymbol{h}_k^{\natural} \boldsymbol{x}_k^{\natural *} \boldsymbol{a}_{kj} + e_j,\ 1 \leq j \leq m, \tag{4.2}$$

which resembles the blind demixing model defined in (3.13) as presented in Sect. 3.2.3. From the observation y_j for $1 \leq j \leq m$, the active set $\mathscr{S}$, data information $\{\boldsymbol{x}_k\}$, and the channel state information $\{\boldsymbol{h}\}$ can be recovered. Hence, joint device activity detection, data decoding, and channel estimation can be achieved.

4.2.2 Multiple-Antenna Scenario

Considering an IoT network consisting of a BS equipped with r antennas and s single-antenna devices with sporadic traffic. Denote $\boldsymbol{g}_{ij}^{\natural} \in \mathbb{C}^m$ as the channel impulse response from the j-th device to the i-th antenna of the BS and recall the transmitted signal at the j-th device defined in (3.6). Thus, the observations $z_i \in \mathbb{C}^m$ at the i-th antenna of the BS are represented as

$$z_i = \sum_{j\in\mathscr{S}} \boldsymbol{f}_j^{\natural} * \boldsymbol{g}_{ij}^{\natural} + \boldsymbol{n}_i,\ \forall i = 1 \ldots r, \tag{4.3}$$

where $\boldsymbol{n}_i \in \mathbb{C}^m$ is additive white complex Gaussian noise. The sparse blind demixing model in the single-antenna scenario is the specific case of (4.3) when $r = 1$. Given the observations $\{z_i\}$, our goal is to detect the active device set $\mathscr{S}$ and recover the associated $\{\boldsymbol{f}_j^{\natural}\}$ and $\{\boldsymbol{g}_{ij}^{\natural}\}$ simultaneously.

Similar to the model in the single-antenna scenario, i.e., (4.2), the l-th entry of $\boldsymbol{y}_i$ is given by

$$\boldsymbol{y}_i[l] = \sum_{j\in\mathscr{S}} \boldsymbol{b}_l^{\mathsf{H}} \boldsymbol{h}_{ij}^{\natural} \boldsymbol{x}_j^{\natural\mathsf{H}} \boldsymbol{c}_{jl} + \boldsymbol{\xi}_i[l],\ l = 1, \ldots, m,\ i = 1, \ldots, r. \tag{4.4}$$

The goal is to simultaneously detect active device set $\mathscr{S}$ and recover both $\{\boldsymbol{x}_j^{\natural}\}$ and $\{\boldsymbol{h}_{ij}^{\natural}\}$ from the observations $\{z_i\}$.

4.3 Convex Relaxation Approach

In this section, we present a convex relaxation approach to solve the sparse blind demixing problem. Taking the single-antenna scenario as an example, the optimization problem is firstly established for the sparse blind demixing model (4.2). Then a convex relaxation approach is further presented to solve this resulting nonconvex optimization problem.

Define a collection of groups as

$$\mathcal{G} = \{\mathcal{G}_1, \mathcal{G}_2, \ldots, \mathcal{G}_s\} \tag{4.5}$$

with

$$\mathcal{G}_k = \{N(k-1)+1, \ldots, Nk\}$$

and

$$\mathcal{G}_i \cap \mathcal{G}_j = \emptyset$$

for $i \neq j$, and denote an aggregative vector as

$$\boldsymbol{x} = [\boldsymbol{x}_1^\top, \ldots, \boldsymbol{x}_s^\top]^\top \in \mathbb{C}^{Ns},$$

where the index set is

$$\mathcal{V} = \{1, 2, \ldots, Ns\}.$$

With the support of the data vector defined as

$$\text{Supp}(\boldsymbol{x}) = \{i | x_i \neq 0, \forall i \in \mathcal{V}\},$$

the sparse blind demixing problem can be formulated as

$$\begin{aligned} \underset{\{\boldsymbol{x}_k\},\{\boldsymbol{h}_k\}}{\text{minimize}} \quad & \sum_{k=1}^{s} \mathbb{I}(\text{Supp}(\boldsymbol{x}) \cap \mathcal{G}_k \neq \emptyset) \\ \text{subject to} \quad & \sum_{j=1}^{m} \left| \sum_{k=1}^{s} \boldsymbol{b}_j^{\mathsf{H}} \boldsymbol{h}_k \boldsymbol{x}_k^{\mathsf{H}} \boldsymbol{a}_{kj} - y_j \right|^2 \leq \epsilon, \end{aligned} \tag{4.6}$$

where parameter $\epsilon > 0$ is known a priori. Denoting $\boldsymbol{x}^\star$ as a solution of problem (4.6), the set of active devices is given as

$$\mathcal{S}^\star = \{k : \text{Supp}(\boldsymbol{x}) \cap \mathcal{G}_k \neq \emptyset\}.$$

Due to the nonconvex bilinear constraint and the combinatorial objective function, problem (4.6) is highly intractable, which motivates to develop efficient algorithms with good performance.

A natural way is to lift the bilinear model into the linear model with a low-rank matrix [9], i.e.,

$$\boldsymbol{b}_j^{\mathsf{H}}\boldsymbol{h}_k\boldsymbol{x}_k^{\mathsf{H}}\boldsymbol{a}_{kj} = \boldsymbol{b}_j^{\mathsf{H}}\boldsymbol{W}_k\boldsymbol{a}_{kj} \tag{4.7}$$

with $\boldsymbol{W}_k \in \mathbb{C}^{K\times N}$ and

$$\mathrm{rank}(\boldsymbol{W}_k) = 1, \forall k = 1, \ldots, s.$$

The natural idea is to exploit a convex relaxation method to deal with the sparsity and low-rankness in matrices $\boldsymbol{W}_k$'s of problem (4.6):

$$\begin{aligned} \underset{\{\boldsymbol{W}_k\}}{\text{minimize}} \quad & \lambda_1 \sum\nolimits_{k=1}^{s} \|\boldsymbol{W}_k\|_* + \lambda_2 \sum\nolimits_{k=1}^{s} \|\boldsymbol{W}_k\|_F \\ \text{subject to} \quad & \sum\nolimits_{j=1}^{m} \left| \sum\nolimits_{k=1}^{s} \boldsymbol{b}_j^{\mathsf{H}}\boldsymbol{W}_k\boldsymbol{a}_{kj} - y_j \right|^2 \le \epsilon, \end{aligned} \tag{4.8}$$

where $\lambda_1 \ge 0$ and $\lambda_2 \ge 0$ are the regularization parameters. The group sparsity structure in the aggregated data signals $\boldsymbol{x}$ induces a group sparsity structure in the lifting vector

$$\mathrm{vec}(\boldsymbol{W}) = [\mathrm{vec}(\boldsymbol{W}_1)^{\mathsf{H}}, \ldots, \mathrm{vec}(\boldsymbol{W}_s)^{\mathsf{H}}]^{\mathsf{H}} \in \mathbb{C}^{KNs},$$

where $\mathrm{vec}(\boldsymbol{M})$ is the vectorization of matrix $\boldsymbol{M}$. Furthermore, the ℓ_1/ℓ_2-norm is adopted to induce the group sparsity in the vector $\mathrm{vec}(\boldsymbol{W})$, i.e.,

$$\|\mathrm{vec}(\boldsymbol{W})\|_{1,2} = \sum_{k=1}^{s} \|\mathrm{vec}(\boldsymbol{W}_k)\|_2 = \sum\nolimits_{k=1}^{s} \|\boldsymbol{W}_k\|_F .$$

4.4 Difference-of-Convex-Functions (DC) Programming Approach

Although the convex relaxation approach (4.8) provides a natural way to solve problem (4.6), the results obtained from norm relaxation are usually suboptimal to the original nonconvex optimization problem [10]. Moreover, two regularization parameters are introduced by the combination of norms, which are difficult to tune. Additionally, there is no efficient convex relaxation approach to simultaneously induce low-rankness and sparsity [2]. To address these issues, the paper [6] developed a difference-of-convex-functions (DC) representation for the rank function in order to satisfy the fixed-rank constraint.

In the sequel, we consider the sparse blind demixing model under the multiple-antenna BS scenario. Specifically, the sparse blind demixing problem is reformulated as a sparse and low-rank matrix recovery problem via lifting the bilinear model into the linear model. Based on the linear model, an exact DC formulation for the rank constraint is further established, followed by developing an efficient DC algorithm (DCA) for minimizing the DC objective.

4.4.1 Sparse and Low-Rank Optimization

The estimation problem for sparse blind demixing with a multiple-antenna BS can be established in the similar form of the optimization problem (4.6). To facilitate the design of the DC algorithm, a sparse and low-rank optimization problem is first established. Denote

$$\boldsymbol{h}_j^{\natural} = \left[\boldsymbol{h}_{1j}^{\natural\mathsf{H}}, \ldots, \boldsymbol{h}_{rj}^{\natural\mathsf{H}}\right]^{\mathsf{H}}, \forall j = 1, \ldots, s, \tag{4.9}$$

where $\boldsymbol{h}_j^{\natural} \in \mathbb{C}^{rk}$. Define a set of matrices

$$\boldsymbol{X}_{ij}^{\natural} = \boldsymbol{h}_{ij}^{\natural}\boldsymbol{x}_j^{\natural\mathsf{H}}, \tag{4.10}$$

where $\boldsymbol{X}_{ij}^{\natural} \in \mathbb{C}^{k\times d}$. Here, $\boldsymbol{X}_{ij}^{\natural}$ is a rank-one matrix when $j \in \mathscr{S}$, otherwise a zero matrix. Define

$$\boldsymbol{X}_j^{\natural} = \left[\boldsymbol{X}_{1j}^{\natural\mathsf{H}}, \boldsymbol{X}_{2j}^{\natural\mathsf{H}}, \ldots, \boldsymbol{X}_{rj}^{\natural\mathsf{H}}\right]^{\mathsf{H}} = \boldsymbol{h}_j^{\natural}\boldsymbol{x}_j^{\natural\mathsf{H}}, \tag{4.11}$$

where $\boldsymbol{X}_j^{\natural} \in \mathbb{C}^{rk\times d}$. $\boldsymbol{X}_j^{\natural}$ is a rank-one matrix when $j \in \mathscr{S}$; otherwise it is a zero matrix. With a matrix defined as $\boldsymbol{E}_i \in \mathbb{R}^{k\times rk}$

$$\boldsymbol{E}_i = \left[\boldsymbol{e}_{k(i-1)+1}, \boldsymbol{e}_{k(i-1)+2}, \ldots, \boldsymbol{e}_{ki}\right]^{\mathsf{H}}, \forall i = 1, \ldots, r, \tag{4.12}$$

where $\boldsymbol{e}_l$ denotes the rk-dimensional standard basis vector, a linear map $\mathscr{A}_{ij} : \mathbb{C}^{rk\times d} \to \mathbb{C}^m$ for $1 \leq i \leq r, 1 \leq j \leq s$ is given by

$$\mathscr{A}_{ij}(\boldsymbol{Z}) := \{\langle \boldsymbol{b}_l \boldsymbol{c}_{jl}^{\mathsf{H}}, \boldsymbol{E}_i\boldsymbol{Z}\rangle\}_{l=1}^m, \tag{4.13}$$

where $\boldsymbol{Z} \in \mathbb{C}^{rk\times d}$ and $\boldsymbol{E}_i\boldsymbol{X}_j^{\natural} = \boldsymbol{X}_{ij}^{\natural}$. Thus, the model (4.4) can be transformed into

$$\boldsymbol{y}_i = \sum_{j=1}^{s} \mathscr{A}_{ij}(\boldsymbol{X}_j^{\natural}) + \boldsymbol{\xi}_i, \forall i = 1, \ldots, r. \tag{4.14}$$

The measurements $\{y_i\}$ are the linear combinations of the corresponding entries of every column block in the lifted matrix $X^\natural \in \mathbb{C}^{rk\times ds}$, which is given by

$$X^\natural = \begin{bmatrix} h_{11}^\natural x_1^{\natural\mathsf{H}} & h_{12}^\natural x_2^{\natural\mathsf{H}} & \dots & h_{1s}^\natural x_s^{\natural\mathsf{H}} \\ h_{21}^\natural x_1^{\natural\mathsf{H}} & h_{22}^\natural x_2^{\natural\mathsf{H}} & \dots & h_{2s}^\natural x_s^{\natural\mathsf{H}} \\ \vdots & \vdots & \ddots & \vdots \\ h_{r1}^\natural x_1^{\natural\mathsf{H}} & h_{r2}^\natural x_2^{\natural\mathsf{H}} & \dots & h_{rs}^\natural x_s^{\natural\mathsf{H}} \end{bmatrix} = \left[X_1^\natural, \dots, X_s^\natural\right].$$

Instead of recovering both $\{h_{ij}^\natural\}$ and $\{x_j^\natural\}$, problem $\mathscr{P}$ is solved with the recovery of the matrix $X^\natural$. Notice that $X^\natural$ has block-low-rank and column sparse structures. The goal is to recover $X^\natural$ from the observation y_i for $i = 1, \dots, r$. Since $X^\natural$ has block-low-rank and column sparse structures, we can establish a sparse and low-rank optimization problem as follows:

$$\begin{aligned} \underset{\{X_j\}}{\text{minimize}} \quad & \left\| [\|\text{vec}(X_1)\|_2, \dots, \|\text{vec}(X_j)\|_2] \right\|_0 \\ \text{subject to} \quad & \sum_{i=1}^{r} \left\| y_i - \sum_{j=1}^{s} \mathscr{A}_{ij}(X_j) \right\|_2^2 \le \epsilon \\ & \text{rank}(X_j) \le 1, \ \forall j = 1, \dots, s, \end{aligned} \tag{4.15}$$

where $\{X_j\} \in \mathbb{C}^{rk\times d}$.

4.4.2 A DC Formulation for Rank Constraint

Before giving an exact DC formulation for the rank constraint, we introduce the definition of Ky Fan k-norm.

Definition 4.1 Ky Fan k-norm [7]: the Ky Fan k-norm of a matrix $X \in \mathbb{C}^{m\times n}$ is defined as the sum of its largest-k singular values, i.e.,

$$|||X|||_k = \sum_{i=1}^{k} \sigma_i(X), \tag{4.16}$$

where $k \le \min\{m, n\}$.

Since the rank of a matrix is equal to the number of its nonzero singular values, for any matrix $\boldsymbol{X} \in \mathbb{C}^{m\times n}$ whose rank is less than k, it can yield from Definition 4.1 that [7]:

$$\operatorname{rank}(\boldsymbol{X}) \leq k \Leftrightarrow \|\boldsymbol{X}\|_* - |||\boldsymbol{X}|||_k = 0. \tag{4.17}$$

Instead of using the discontinuous rank function, a continuous DC function $\|\boldsymbol{X}\|_* - |||\boldsymbol{X}|||_k$ can be adopted for inducing low-rankness property of a matrix.

4.4.3 DC Algorithm for Minimizing a DC Objective

Based on (4.17), problem (4.15) can be further formulated as the minimization problem with a DC objective function:

$$\begin{aligned} \underset{\{\boldsymbol{X}_j\}}{\text{minimize}} \quad & \sum_{j=1}^{s} \left(\|\boldsymbol{X}_j\|_* - \left|\left|\left|\boldsymbol{X}_j\right|\right|\right|_1 \right) \\ \text{subject to} \quad & \sum_{i=1}^{r} \left\| \boldsymbol{y}_i - \sum_{j=1}^{s} \mathscr{A}_{ij}(\boldsymbol{X}_j) \right\|_2^2 \leq \epsilon. \end{aligned} \tag{4.18}$$

To address the nonconvexity of the DC objective function, a DC algorithm based on majorization-minimization (MM) has been proposed in [12]. At each iteration, the DC algorithm solves a convex subproblem, given by

$$\begin{aligned} \underset{\{\boldsymbol{X}_j\}}{\text{minimize}} \quad & \sum_{j=1}^{s} \left(\|\boldsymbol{X}_j\|_* - \langle \boldsymbol{X}_j, \boldsymbol{Y}_j^{t-1} \rangle \right) \\ \text{subject to} \quad & \sum_{i=1}^{r} \left\| \boldsymbol{y}_i - \sum_{j=1}^{s} \mathscr{A}_{ij}(\boldsymbol{X}_j) \right\|_2^2 \leq \epsilon, \end{aligned} \tag{4.19}$$

where $\boldsymbol{Y}_j^{t-1} \in \mathbb{C}^{rk\times d}$ is a subgradient of $\left|\left|\left|\boldsymbol{X}_j\right|\right|\right|_1$ at $\boldsymbol{X}_j^{t-1}$ and can be efficiently derived from the singular value decomposition, given by

$$\partial \left|\left|\left|\boldsymbol{X}_j^t\right|\right|\right|_1 = \left\{ \boldsymbol{U} \operatorname{diag}(\boldsymbol{q}) \boldsymbol{V}^H \ : \boldsymbol{q} = [1, 0, \ldots, 0] \right\}. \tag{4.20}$$

The DC algorithm is illustrated in Algorithm 4.1.

Algorithm 4.1: DC algorithm for problem (4.18)

Input: $\{\mathscr{A}_{ij}\}$, $\{y_i\}$, upper bound ϵ, a small value η
Output: $\{X_j^t\}$

Initialisation : $\{X_j^0\}$
1: $k = 1$
LOOP Process
2: for $t = 1, 2, \ldots$ do
3: Select $\{Y_j^{t-1} \in \partial \|X_j^{t-1}\|_k\}$
4: Solve the convex problem (4.19), and obtain the optimal solution $\{X_j^t\}$
5: if $\sum_{j=1}^{s}(\|X_j^t\|_* - |||X_j^t|||_1) < \eta$ then
6: break
7: end if
8: end for
9: return $\{X_j^t\}$

4.4.4 Simulations

In this section, we conduct numerical experiments to compare the proposed DC approach with the convex relaxation methods for empirical recovery performance and test the robustness against noise.

For $j \notin \mathscr{S}$, set the ground truth data signal as $x_j^\natural = \mathbf{0}_d$, and for $j \in \mathscr{S}$, $x_j^\natural$ is drawn i.i.d. from the standard complex Gaussian distribution. Both the channel states $\{h_{ij}^\natural\}$ and matrices $\{C_j\}$ are drawn i.i.d. from the standard complex Gaussian distribution. To measure the accuracy of estimation, the relative construction error is defined as

$$\operatorname{error}(X) = \frac{\sqrt{\sum_{j=1}^{s}\|X_j^\natural - X_j\|_F^2}}{\sqrt{\sum_{j=1}^{s}\|X_j^\natural\|_F^2}}, \tag{4.21}$$

where the ground truth matrices are denoted as $\{X_j^\natural\}$, and $\{X_j\}$ are the estimated matrices.

We compare the empirical recovery and robustness performance of the following four algorithms:

- **DC algorithm (DCA)**: The termination criterion is either the iteration number exceeding 200 or $\sum_{j=1}^{s}(\|X_j^t\|_* - |||X_j^t|||_1) < 10^{-6}$.
- **mixed norm minimization(MNM)**: The regularization terms λ_1 and λ_1 are chosen via cross validation.

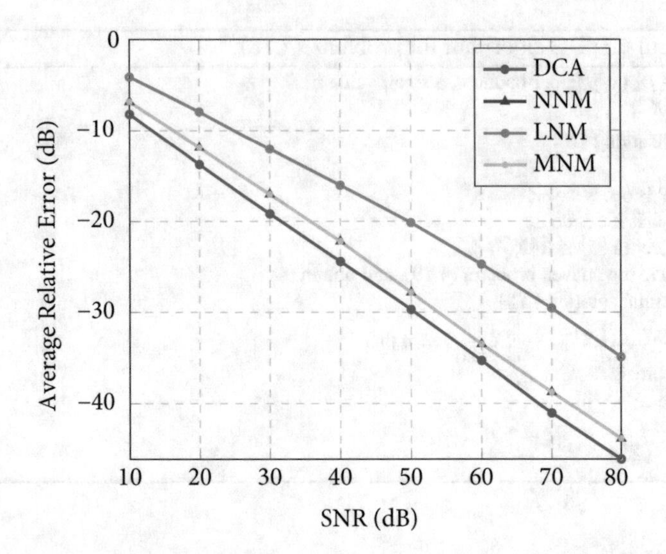

Fig. 4.1 Probability of successful recovery with different sample sizes m

- **nuclear norm minimization (NNM)**: The algorithm is similar to **MNM** except for $\lambda_1 = 1, \lambda_2 = 0$.
- ℓ_1/ℓ_2**-norm minimization(LNM)**: The algorithm is similar to **MNM** except for $\lambda_1 = 0, \lambda_2 = 1$.

The empirical recovery performance of the above four algorithms is investigated in the noiseless scenario under the setting of $k = 5, d = 20, s = 10, |\mathscr{S}| = 4$, and $r = 3$. For each setting, 20 independent trails are performed and the recovery is regarded as a success if the error($\boldsymbol{X}$) $< 10^{-2}$. Figure 4.1 shows the performance of recovery with varying the number of measurements.

The robustness of the four algorithms with respect to noise is further investigated. The noise $\boldsymbol{\xi}_i$ is generated as

$$\boldsymbol{\xi}_i = \sigma \cdot \|\boldsymbol{y}_i\| \cdot \frac{z_i}{\|z_i\|}, \ \forall i = 1 \ldots r, \tag{4.22}$$

where $z_i \in \mathbb{C}^m$ is the normalized standard Gaussian vector. Under the setting of $k = 5, d = 20, s = 10, |\mathscr{S}| = 4, r = 3$, and $m = 670$, 20 independent trails are performed with respect to different σ. Figure 4.2 illustrates the average relative error in dB against the signal-to-noise-ratio (SNR). It shows that DCA enjoys a higher accuracy of reconstruction than other algorithms.

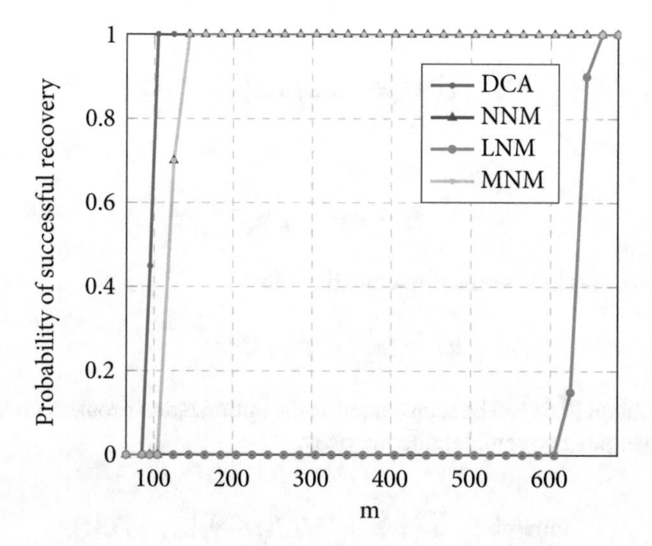

Fig. 4.2 Robustness under different SNR(dB)

4.5 Smoothed Riemannian Optimization on Product Manifolds

Another line of literatures have developed efficient nonconvex algorithms to solve the sparse and low-rank optimization problem [8, 13] in the natural vector space via matrix factorization. For instance, an alternating minimization approach was developed in [8] for solving the sparse blind deconvolution problem. However, the additional group sparsity structure of the sparse blind demixing problem (4.6) brings unique challenges to develop the nonconvex optimization paradigm. To address this challenge, a smoothed Riemannian optimization approach is introduced to solve sparse blind demixing problem [4], thereby achieving better performance with low computational complexity. More details on the manifold optimization can be referred to Sect. 3.4.3.

4.5.1 Optimization on Product Manifolds

To begin with, problem (4.6) is formulated as a regularized optimization problem under fixed-rank constraints. For $k = 1, \ldots, s$, $j = 1, \ldots, m$, define

$$\boldsymbol{c}_j = [\boldsymbol{b}_j^{\mathsf{H}}, \boldsymbol{0}_N^{\mathsf{H}}]^{\mathsf{H}} \in \mathbb{C}^{N+K}, \quad \boldsymbol{d}_{kj} = [\boldsymbol{0}_K^{\mathsf{H}}, \boldsymbol{a}_{kj}^{\mathsf{H}}]^{\mathsf{H}} \in \mathbb{C}^{N+K},$$

it yields

$$c_j^{\mathsf{H}} M_k d_{kj} = b_j^{\mathsf{H}} h_k x_k^{\mathsf{H}} a_{kj}, \tag{4.23}$$

where

$$M_k = w_k w_k^{\mathsf{H}} \in \mathbb{S}_+^{N+K} \tag{4.24}$$

is a Hermitian positive semidefinite matrix with

$$w_k = [h_k^{\mathsf{H}}, x_k^{\mathsf{H}}]^{\mathsf{H}} \in \mathbb{C}^{N+K}. \tag{4.25}$$

Hence, problem (4.6) can be represented as the optimization problem on the product of Hermitian positive semidefinite matrices:

$$\begin{aligned} &\underset{M}{\text{minimize}} \quad \sum_{j=1}^{m} \Big| \sum_{k=1}^{s} c_j^{\mathsf{H}} M_k d_{kj} - y_j \Big|^2 + \lambda f(M) \\ &\text{subject to} \quad \text{rank}(M_k) = 1,\ k = 1, \ldots, s, \end{aligned} \tag{4.26}$$

where $M = \{M_k\}_{k=1}^{s}$ with $M_k \in \mathbb{S}_+^{N+K}$, $\lambda > 0$ is the regularization parameter, and $f(M)$ is the function to induce the sparsity structure. Here, M_k is in the space of the manifold encoded by complex symmetric rank-one matrices, i.e., $M_k \in \mathcal{M}_k$ [5]. It yields that $M \in \mathcal{M}^s$, where

$$\mathcal{M}^s := \mathcal{M}_1 \times \mathcal{M}_2 \times \cdots \times \mathcal{M}_s \tag{4.27}$$

represents the product of manifolds $\mathcal{M}_k$. By exploiting the quotient manifold geometry of the product of complex symmetric rank-one matrices, computationally efficient Riemannian optimization algorithms can be developed on product manifolds.

4.5.2 Smoothed Riemannian Optimization

The smooth objective function is normally required [1, 11] in order to develop Riemannian optimization algorithm for solving problem (4.26). To achieve this goal, the smoothed ℓ_1/ℓ_2-norm is introduced, represented as

$$f_\epsilon(M) = \sum\nolimits_{k=1}^{s} \left(\|M_k\|_F^2 + \epsilon^2 \right)^{1/2} \tag{4.28}$$

with $\epsilon > 0$ as the smoothing parameter with a small value. This can be used for inducing the group sparsity structure in vector

$$\mathrm{vec}(\boldsymbol{M}) = [\mathrm{vec}(\boldsymbol{M}_1)^{\mathsf{H}}, \ldots, \mathrm{vec}(\boldsymbol{M}_s)^{\mathsf{H}}]^{\mathsf{H}}. \tag{4.29}$$

Therefore, the proposed smoothed Riemannian optimization approach over the product manifold $\mathscr{M}^s$ for sparse blind demixing problem (4.6) is given by

$$\underset{\boldsymbol{M} \in \mathscr{M}^s}{\text{minimize}} \sum_{j=1}^{m} \Big| \sum_{k=1}^{s} \boldsymbol{c}_j^{\mathsf{H}} \boldsymbol{M}_k \boldsymbol{d}_{kj} - y_j \Big|^2 + \lambda f_\epsilon(\boldsymbol{M}), \tag{4.30}$$

where the objective function is smooth and the constraint is a manifold.

Due to the geometry of the product manifolds, the Riemannian optimization algorithms developed on the product manifold $\mathscr{M}^s$ can be elementwisely operated over the individual manifolds $\mathscr{M}_k$ [5]. For individual manifold $\mathscr{M}_k$, the descent direction is detected on the horizontal space of the manifold and then retract it on the manifold via retraction operation. Therein, the detection of the descent direction can be achieved by the Riemannian optimization algorithms, e.g., conjugate gradient descent algorithm [1].

4.5.3 Simulation Results

In this section, to illustrate the advantages of the smoothed Riemannian optimization for solving the sparse blind demixing problem (4.30), the Riemannian conjugate-gradient descent algorithm (RCGD) is compared with the other three algorithms mentioned in Sect. 4.4.4, i.e., nuclear norm minimization (NNM), ℓ_1/ℓ_2-norm minimization (LNM), and mixed norm minimization (MNM). Here, the RCGD algorithm adopts the initialization strategy in [5] and stops when the norm of Riemannian gradient falls below 10^{-8} or the number of iterations exceeds 500.

The empirical recovery performance of the above four algorithms, i.e., RCGD, NNM, LMN, and MNM, are investigated under the setting of $N = K = 10$, $s = 10$, $|\mathscr{A}| = 3$. For each setting, 30 independent trails are performed and the recovery is considered as a success if $\mathrm{err}(\boldsymbol{x}) \leq 10^{-2}$. Figure 4.3 illustrates the probability of successful recovery with respect to different sample sizes m. It shows that the smoothed Riemannian optimization algorithm achieves much better performance than other algorithms. That is, it exactly recovers the ground truth signals with less samples.

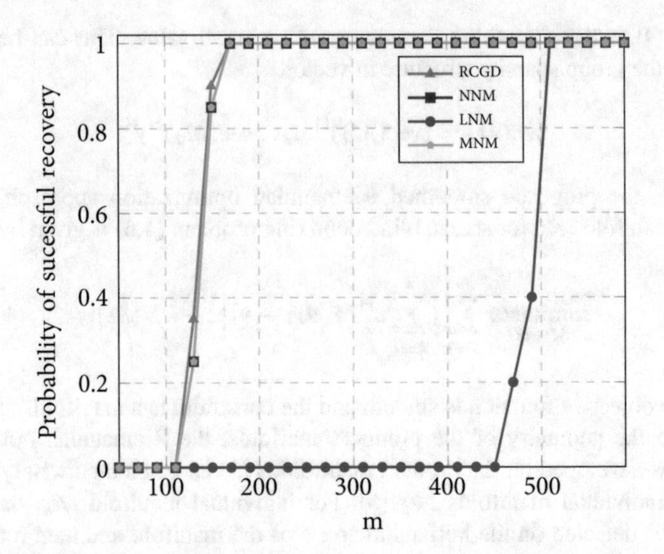

Fig. 4.3 Probability of successful recovery with different sample sizes m

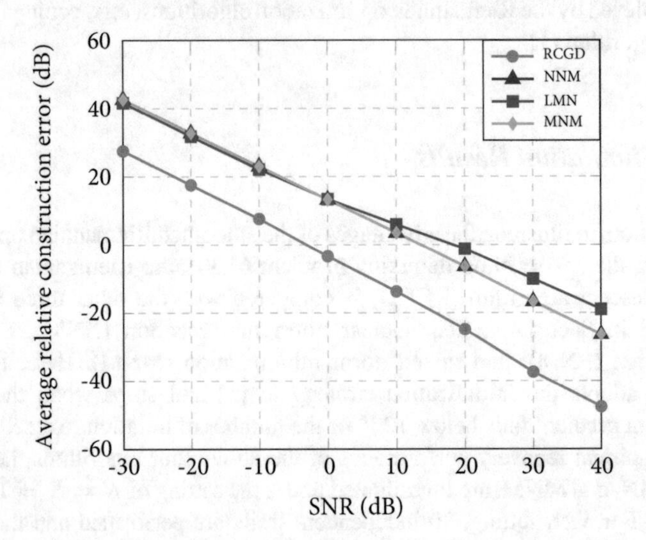

Fig. 4.4 Average relative construction error vs. SNR (dB)

The average relative construction error of the four algorithms is further investigated to explore the robustness of the proposed smoothed Riemannian optimization algorithm against additive noise. The four algorithms for each level of signal-to-noise ratio (SNR) $1/\sigma$ are compared in the setting of $m = 550$, $N = K = 10$, $s = 10$, $|\mathscr{A}| = 3$. For each setting, 20 independent trails are performed. The average relative construction error in dB against the SNR is showed in Fig. 4.4, which demonstrates that RCGD is robust to the noise and can achieve better performance than other algorithms.

4.6 Summary

This chapter introduced a sparse blind demixing model with both single-antenna and multi-antenna BSs for joint device activity detection, data decoding, and channel estimation in IoT networks with the grant-free random access scheme. It enjoys attractive advantages by removing the overhead caused by channel estimation sequence and device activity information. According to the simultaneous group sparse and low-rank variables in the sparse blind demixing model, the convex relaxation approach based on the norm minimization was first introduced. To further pursue higher accuracy of signal reconstruction compared to the convex relaxation approach, the approach that minimizes the difference-of-convex (DC) objective functions was developed. Another line of works has been focused on establishing Riemannian manifold to characterize the structured variables in the sparse blind demixing model. It is also interesting to further investigate the geometry property, i.e., group sparsity and low-rankness, of the sparse blind demixing model, thereby facilitating to design efficient algorithms with satisfactory performance, i.e., low sample complexity or high accuracy of estimation. A rigorous theoretical analysis on the sparse blind demixing problem is also of interest for future study, to characterize the number of measurements required for exact recovery.

References

1. Absil, P.A., Mahony, R., Sepulchre, R.: Optimization Algorithms on Matrix Manifolds. Princeton University Press, Princeton (2009)
2. Aghasi, A., Bahmani, S., Romberg, J.: A tightest convex envelope heuristic to row sparse and rank one matrices. In: Proceedings of the IEEE Global Conference on Signal and Information Processing (GlobalSIP), p. 627. IEEE, Piscataway (2013)
3. Dong, J., Shi, Y.: Nonconvex demixing from bilinear measurements. IEEE Trans. Signal Process. **66**(19), 5152–5166 (2018)
4. Dong, J., Shi, Y., Ding, Z.: Sparse blind demixing for low-latency signal recovery in massive IoT connectivity. In: Proceedings of the IEEE International Conference on Acoustics Speech Signal Processing (ICASSP), pp. 4764–4768. IEEE, Piscataway (2019)
5. Dong, J., Yang, K., Shi, Y.: Blind demixing for low-latency communication. IEEE Trans. Wireless Commun. **18**(2), 897–911 (2019)
6. Fu, M., Dong, J., Shi, Y.: Sparse blind demixing for low-latency wireless random access with massive connectivity. In: Proceedings of the IEEE Vehicular Technology Conference (VTC), pp. 4764–4768. IEEE, Piscataway (2019)
7. Gotoh, J.Y., Takeda, A., Tono, K.: DC formulations and algorithms for sparse optimization problems. Math. Program. **169**(1), 141–176 (2018)
8. Lee, K., Wu, Y., Bresler, Y.: Near-optimal compressed sensing of a class of sparse low-rank matrices via sparse power factorization. IEEE Trans. Inf. Theory **64**(3), 1666–1698 (2018)
9. Ling, S., Strohmer, T.: Blind deconvolution meets blind demixing: algorithms and performance bounds. IEEE Trans. Inf. Theory **63**(7), 4497–4520 (2017)
10. Lu, C., Tang, J., Yan, S., Lin, Z.: Nonconvex nonsmooth low rank minimization via iteratively reweighted nuclear norm. IEEE Trans. Image Process. **25**(2), 829–839 (2016)

11. Shi, Y., Mishra, B., Chen, W.: Topological interference management with user admission control via Riemannian optimization. IEEE Trans. Wireless Commun. **16**(11), 7362–7375 (2017)
12. Tao, P.D., An, L.T.H.: Convex analysis approach to DC programming: theory, algorithms and applications. Acta Math. Vietnam. **22**(1), 289–355 (1997)
13. Zhang, Y., Kuo, H.W., Wright, J.: Structured local optima in sparse blind deconvolution (2018). Preprint. arXiv: 1806.00338

Chapter 5
Shuffled Linear Regression

Abstract In this chapter, we shall introduce a shuffled linear regression model for joint data decoding and device identification in IoT networks. It is first formulated as a maximum likelihood estimation (MLE) problem. To solve this MLE problem, two algorithms are presented: one is based on sorting, and the other algorithm returns an approximate solution to the MLE problem. Next, theoretical analysis on the shuffled linear regression based on the algebraic-geometric theory is presented. Based on the analysis, an algebraically initialized expectation-maximization algorithm is introduced to solve the problem.

5.1 Joint Data Decoding and Device Identification

In the massive IoT scenario, the device identity information plays a vital role in differential updates, spatial correlation [12], and multi-stage collection [11], for which sensors are used to reconstruct the spatial field. It would take excess time if the identity information has to be collected regularly. Hence, a significant gain in the efficiency of communication procedure can be obtained by excluding the identification information in the header of the packet structure. This yields a joint data decoding and device identification problem at the BS, which may also act as a data fusion center. To achieve this goal, a *shuffled linear regression* has been recently investigated in a line of literature [9, 10, 14, 15] that can be exploited to remove the metadata used for device identification. The shuffled linear regression for identification-free communication is illustrated in Fig. 5.1. Considering a massive sensor network that contains m sensor nodes to capture the parameter data $\boldsymbol{x} \in \mathbb{R}^n$ generated from n devices, a shuffled linear regression can be represented as

$$\boldsymbol{y} = \boldsymbol{\Pi} \boldsymbol{A} \boldsymbol{x}, \tag{5.1}$$

where $\boldsymbol{y} \in \mathbb{R}^m$ is the permuted signal received at the BS, $\boldsymbol{A} \in \mathbb{R}^{m \times n}$ is an encoding matrix, and $\boldsymbol{\Pi}$ is an unknown $m \times m$ permutation matrix whose i-th row is the canonical vector $\boldsymbol{e}_{\pi(i)}^{\top}$ of all zeros except a 1 at position $\pi(i)$. The recovery of the shuffled linear regression (5.1) enables the BS to decode the signal

Y. Shi et al., *Low-overhead Communications in IoT Networks*,
https://doi.org/10.1007/978-981-15-3870-4_5

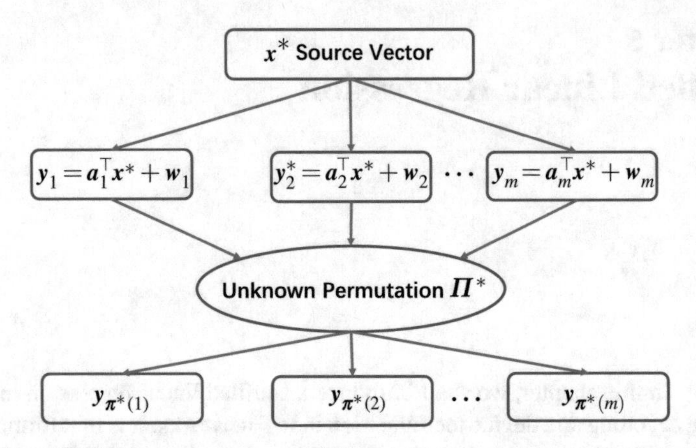

Fig. 5.1 An example to illustrate the shuffled linear regression for identification-free communication

$x = [x_1, \ldots, x_n]^\top$ corresponding to each device from subsampled and permuted measurements y. A simple linear shuffled model is illustrated in Example 5.1.

Example 5.1 Consider a senor network with three sensor nodes and two devices. Assume that the parameter data x and the encoding matrix are given by

$$A = \begin{bmatrix} 1 & 2 \\ -2 & 4 \\ 0 & -5 \end{bmatrix} \quad \text{and} \quad x = \begin{bmatrix} 3 \\ 4 \end{bmatrix}. \tag{5.2}$$

Based on a permutation matrix Π, i.e.,

$$\Pi = \begin{bmatrix} 0 & 0 & 1 \\ 1 & 0 & 0 \\ 0 & 1 & 0 \end{bmatrix} \tag{5.3}$$

it yields a shuffled linear model:

$$y = \Pi \cdot Ax = \begin{bmatrix} 0 & 0 & 1 \\ 1 & 0 & 0 \\ 0 & 1 & 0 \end{bmatrix} \begin{bmatrix} 11 \\ 10 \\ -20 \end{bmatrix} = \begin{bmatrix} -20 \\ 11 \\ 10 \end{bmatrix}. \tag{5.4}$$

Recently, important theoretical advances have been made to understand this problem, which can be mainly separated into three types: statistical approaches, algebraic geometry approaches, and alternating minimization approaches. For statistical approaches, the works [2, 4, 13] have developed algorithms based on the maximum likelihood estimator $\boldsymbol{x}_{\mathrm{ML}}$ given by

$$(\boldsymbol{\Pi}_{\mathrm{ML}}, \boldsymbol{x}_{\mathrm{ML}}) = \underset{\boldsymbol{\Pi}^*, \boldsymbol{x}^*}{\operatorname{argmin}} \left\| \boldsymbol{y} - \boldsymbol{\Pi}^* \boldsymbol{A} \boldsymbol{x}^* \right\|_2.$$

In addition, the conditions when the estimator recovers the ground truth in (5.1), i.e., $\boldsymbol{\Pi}_{\mathrm{ML}} = \boldsymbol{\Pi}$, have been established in the works [9, 13].

When the ratio of shuffled entries to all of the data entries is small, one may apply alternating minimization or multi-start gradient descent to solve (5.9) [2], which is an NP-hard problem for $n > 1$ [15]. Due to high nonconvexity, such methods are very sensitive to initialization. This issue is addressed by the algebraically initialized expectation-maximization method proposed in [15], which uses the solution to the polynomial system of equations mentioned above to obtain a high-quality initialization. This approach is robust to small levels of noise, efficient for $n \leq 5$, and is able to handle fully shuffled data.

In the following, we will first demonstrate that the shuffled linear regression provides a way to achieve joint data decoding and device identification. Furthermore, two types of methods for solving the estimation problem in the shuffled linear regression will be introduced along with theoretical analysis, which include a maximum likelihood estimation based approach and an algebraic-geometric approach.

5.2 Problem Formulation

Consider a massive sensor network that contains m sensor nodes. Based on the correspondence pairs $\{\boldsymbol{u}_j, y_j\}_{j=1}^m$, the aim is to find the parameter vector

$$\boldsymbol{x} = [x_1, \ldots, x_n]^\top \in \mathbb{R}^n$$

that characterizes the environment information, e.g., temperature, humidity, and pressure. The measurements are given by

$$y_j = \boldsymbol{a}_j^\top \boldsymbol{x}, \quad \forall j = 1, \ldots, m, \tag{5.5}$$

$$\boldsymbol{a}_j := [a_1(\boldsymbol{u}_j), \ldots, a_n(\boldsymbol{u}_j)]^\top, \tag{5.6}$$

where $a_i : \mathbb{R}^s \to \mathbb{R}$ are known functions. The shuffled linear regression (5.5) can be considered as a special type of data corruption where the correspondences are missing. It can support identification-free communications, where the identification information, i.e., correspondences in the model (5.5), is excluded from the packet structure.

Given functions of the input samples represented in (5.6), i.e.,

$$\boldsymbol{A} = [\boldsymbol{a}_1, \ldots, \boldsymbol{a}_m]^\top \in \mathbb{R}^{m \times n}, \tag{5.7}$$

and a shuffled signal $\boldsymbol{y} = [y_{j_1}, \ldots, y_{j_m}]^\top \in \mathbb{R}^m$ with the unknown shuffling indices $j_1, \ldots, j_m$, we have

$$\boldsymbol{y} = (\boldsymbol{\Pi})^\top \boldsymbol{A}\boldsymbol{x} + \boldsymbol{w} \in \mathbb{R}^m, \tag{5.8}$$

where $\boldsymbol{x} \in \mathbb{R}^n$, $\boldsymbol{\Pi}$ is an $m \times m$ permutation matrix, and the vector

$$\boldsymbol{w} \sim \mathcal{N}(\boldsymbol{0}, \sigma^2 \boldsymbol{I}_m)$$

represents the additive Gaussian noise. The goal of the shuffled linear regression is to efficiently estimate both the signal $\boldsymbol{x}$ and the permutation matrix $\boldsymbol{\Pi}$ from $\boldsymbol{y}$. Thus, joint data decoding and device identification in the IoT network is achieved.

5.3 Maximum Likelihood Estimation Based Approaches

Several methods have recently been developed to solve the shuffled linear regression problem. Specifically, the estimation of shuffled linear regression can be achieved via the *maximum likelihood estimator* (MLE) [2, 9, 13]:

$$(\widehat{\boldsymbol{\Pi}}_{\text{ML}}, \widehat{\boldsymbol{x}}_{\text{ML}}) = \underset{\boldsymbol{\Pi}^*, \boldsymbol{x}^*}{\operatorname{argmin}} \left\| \boldsymbol{\Pi}^* \boldsymbol{y} - \boldsymbol{A}\boldsymbol{x}^* \right\|_2. \tag{5.9}$$

Based on this estimator, several algorithms have been developed and the theoretical analyses have been established, e.g., [2, 4, 16]. It shows that $\widehat{\boldsymbol{\Pi}}_{\text{ML}}$ is divergent from $\boldsymbol{\Pi}$ in (5.8) with high probability if the SNR is not large enough [4]. The detailed guarantees will be discussed in Theorems 5.1 and 5.2. Furthermore, [16] shows that if the SNR approaches infinity, $\widehat{\boldsymbol{x}}_{\text{ML}}$ approaches $\boldsymbol{x}$ in (5.8). Moreover, advanced algorithms based on algebraic-geometric approaches proposed recently to address the computational issue will also be introduced.

5.3.1 Sorting Based Algorithms

The paper [9] analyzed the shuffled linear regression problem under the assumption that the entries of the matrix $\boldsymbol{A}$ are drawn i.i.d. from a standard Gaussian distribution. The paper [9] established sharp conditions on the sample size m, dimension n, and SNR, under which $\boldsymbol{\Pi}$ is exactly recoverable. From the computational point of view, the paper [9] demonstrated that the maximum likelihood estimate of $\boldsymbol{\Pi}$ is NP-hard to compute, and it proposed a polynomial-time algorithm based on sorting, which is introduced in the sequel.

Theorems 5.1 and 5.2 in the following provide the statistical properties of the MLE (5.9). Based on the maximum likelihood estimator (5.9), an upper bound on the probability of error of $\widehat{\boldsymbol{\Pi}}_{\mathsf{ML}}$ is provided in the following theorem given by [9] with $c_1, c_2 > 0$.

Theorem 5.1 *For any $n < m$ and $\epsilon < \sqrt{m}$, if*

$$\log\left(\frac{\|\boldsymbol{x}\|_2^2}{\sigma^2}\right) \geq \left(c_1 \frac{m}{m-n} + \epsilon\right) \log m, \tag{5.10}$$

then $\mathbb{P}\{\widehat{\boldsymbol{\Pi}}_{\mathsf{ML}} \neq \boldsymbol{\Pi}\} \leq c_2 m^{-2\epsilon}$.

Furthermore, the lower bound on the probability of error of $\widehat{\boldsymbol{\Pi}}_{\mathsf{ML}}$ is provided as follows.

Theorem 5.2 *For any $\delta \in (0, 2)$, if*

$$2 + \log\left(1 + \frac{\|\boldsymbol{x}\|_2^2}{\sigma^2}\right) \leq (2-\delta) \log m, \tag{5.11}$$

then $\mathbb{P}\{\widehat{\boldsymbol{\Pi}} \neq \boldsymbol{\Pi}\} \geq 1 - c_3 e^{-c_4 m \delta}$ for any estimator $\widehat{\boldsymbol{\Pi}}$.

We can conclude from Theorem 5.2 that if condition (5.11) is satisfied, the recovery probability approaches 1 when m tends to infinity.

Since Eq. (5.9) requires a combinatorial minimization over $n!$ permutations, advanced algorithms are needed to compute $\widehat{\boldsymbol{\Pi}}_{\mathsf{ML}}$ efficiently. To begin with, the maximum likelihood estimate of the permutation is represented as [9]

$$\widehat{\boldsymbol{\Pi}}_{\mathsf{ML}} = \arg\min_{\boldsymbol{\Pi}} \| P_{\boldsymbol{\Pi}}^{\perp} \boldsymbol{y}\|_2^2, \tag{5.12}$$

where

$$P_{\boldsymbol{\Pi}}^{\perp} = \boldsymbol{I} - \boldsymbol{\Pi}\boldsymbol{A}(\boldsymbol{A}^{\top}\boldsymbol{A})^{-1}(\boldsymbol{\Pi}\boldsymbol{A})^{\top}.$$

When $n = 1$ and representing the design vector as $\boldsymbol{a}$, Eq. (5.12) can be represented as

$$\begin{aligned} \widehat{\boldsymbol{\Pi}}_{\mathsf{ML}} &= \arg\max_{\boldsymbol{\Pi}} \|\boldsymbol{a}_{\boldsymbol{\Pi}}^{\top}\boldsymbol{y}\|^2 \\ &= \arg\max_{\boldsymbol{\Pi}} \max\left\{\boldsymbol{a}_{\boldsymbol{\Pi}}^{\top}\boldsymbol{y}, -\boldsymbol{a}_{\boldsymbol{\Pi}}^{\top}\boldsymbol{y}\right\} \\ &= \arg\min_{\boldsymbol{\Pi}} \max\left\{\|\boldsymbol{a}_{\boldsymbol{\Pi}} - \boldsymbol{y}\|_2^2, \|\boldsymbol{a}_{\boldsymbol{\Pi}} + \boldsymbol{y}\|_2^2\right\}. \end{aligned} \tag{5.13}$$

The polynomial-time algorithm illustrated in Algorithm 5.1 is developed based on (5.13). This is achieved based on the fact that for fixed vectors $\boldsymbol{x}$ and $\boldsymbol{y}$,

$$\|\boldsymbol{x}_{\boldsymbol{\Pi}} - \boldsymbol{y}\|$$

can be minimized for $\boldsymbol{\Pi}$ by sorting $\boldsymbol{x}$ according to the order of $\boldsymbol{y}$. The theoretical analysis of Algorithm 5.1 from the computational points of view is illustrated in Theorem 5.3.

Algorithm 5.1: Exact algorithm for implementing Eq. (5.12)

Input: design vector $\boldsymbol{a}$, observation vector $\boldsymbol{y}$
1 $\boldsymbol{\Pi}_1 \leftarrow$ permutation that sorts $\boldsymbol{a}$ according to $\boldsymbol{y}$
2 $\boldsymbol{\Pi}_2 \leftarrow$ permutation that sorts $-\boldsymbol{a}$ according to $\boldsymbol{y}$
3 $\widehat{\boldsymbol{\Pi}}_{\mathrm{ML}} \leftarrow \arg\max\{|\boldsymbol{a}_{\boldsymbol{\Pi}_1}^{\top}\boldsymbol{y}|, |\boldsymbol{a}_{\boldsymbol{\Pi}_2}^{\top}\boldsymbol{y}|\}$
Output: $\widehat{\boldsymbol{\Pi}}_{ML}$

Theorem 5.3 *For $n = 1$, the MLE estimator $\widehat{\boldsymbol{\Pi}}_{\mathrm{ML}}$ can be computed via Algorithm 5.1 in time $\mathcal{O}(m \log m)$ for any choice of the measurement matrix* $\boldsymbol{A}$. *In contrast, if $n > 1$, then $\widehat{\boldsymbol{\Pi}}_{\mathrm{ML}}$ is NP-hard to compute.*

Theorem 5.3 shows that the algorithmic advantages enjoyed in the case of $n = 1$ cannot extend to general cases of $n > 1$. For $n > 1$ a natural method is brute force search: for each permutation $\boldsymbol{\Pi}$ of the $m!$ permutations, check whether the linear system

$$\boldsymbol{\Pi}\boldsymbol{y} = \boldsymbol{A}\boldsymbol{x}$$

is consistent, followed by solving it if it is consistent. This algorithm endows with the complexity of $\mathcal{O}(n^2(m+1)!)$. An approximate algorithm that is more efficient than the brute force has been proposed in [4], which makes progress on both computational and statistical aspects. It is introduced in the next section.

5.3.2 *Approximation Algorithm*

Considering the least squares problem (5.9), an approximation approach [4] is proposed that for any $\epsilon \in (0, 1)$, it returns an $(1 + \epsilon)$-approximation in time $\mathcal{O}((m/\epsilon)^n)$.

The approximation algorithm, shown as Algorithm 5.3, uses a careful enumeration to beat the naive brute force running time of $\Omega(n!)$. The "Row Sampling" algorithm [3] is exploited in the beginning of Algorithm 5.3 in order to narrow the search space. The details of the "Row Sampling" algorithm [3] is presented in the following. The "Row Sampling" is illustrated in Algorithm 5.2 with the following notations:

- For each $i \in [n]$, $\boldsymbol{e}_i$ is the i-th coordinate basis vector in $\mathbb{R}^n$.
- $L(\boldsymbol{x}, \delta_L, \boldsymbol{A}, \ell) := \dfrac{\boldsymbol{x}^\top(\boldsymbol{A} - (\ell + \delta_L)\boldsymbol{I}_k)^{-2}\boldsymbol{x}}{\phi(\ell + \delta_L, \boldsymbol{A}) - \Phi(\ell, \boldsymbol{A})} - (\ell + \delta_L)\boldsymbol{I}_k)^{-1}\boldsymbol{x}$, where $\phi(\ell, \boldsymbol{A}) := \sum_{i=1}^{k} \frac{1}{\lambda_i(\boldsymbol{A}) - \ell}$ and $(\lambda_i(\boldsymbol{A}))_{i=1}^{k}$ are the eigenvalues of $\boldsymbol{A}$.
- $\hat{U}(\boldsymbol{x}, \delta, \boldsymbol{B}, u) := \dfrac{\boldsymbol{x}^\top(\boldsymbol{B} - u'\boldsymbol{I}_r)^{-2}\boldsymbol{x}}{\phi'(u, \boldsymbol{B}) - \phi'(u', \boldsymbol{B})} - \boldsymbol{x}^\top(\boldsymbol{B} - u'\boldsymbol{I}_r)^{-1}\boldsymbol{x}$, where $u' = u + \delta$, $\phi'(u, \boldsymbol{B}) := \sum_{i=1}^{r} \frac{1}{u - \lambda_i(\boldsymbol{B})}$, and $(\lambda_i(\boldsymbol{B}))_{i=1}^{k}$ are the eigenvalues of $\boldsymbol{B}$.

Algorithm 5.2: "Row Sampling" algorithm [3]

---

input Matrix $\boldsymbol{A} = [\boldsymbol{A}_1 \cdots \boldsymbol{A}_n]^\top \in \mathbb{R}^{m \times n}$ such that $\boldsymbol{A}^\top \boldsymbol{A} = \boldsymbol{I}_n$; integer $r \geq n$.
output Matrix $\boldsymbol{S} = (S_{i,j})_{(i,j) \in \times [m]} \in \mathbb{R}^{r \times m}$.
1: Set $\boldsymbol{Q}_0 = \boldsymbol{0}_{n \times n}$, $\boldsymbol{B}_0 = \boldsymbol{0}_{m \times m}$, $\boldsymbol{S} = \boldsymbol{0}_{r \times m}$, $\delta = (1 + m/r)(1 - \sqrt{n/r})^{-1}$ and $\delta_L = 1$.
2: for $\tau = 0$ to $r - 1$ do
3: Let $\ell_\tau = \tau - \sqrt{rk}$ and $u_\tau = \delta(\tau + \sqrt{mr})$.
4: Select $i_\tau \in [m]$ and number $t_\tau > 0$ such that $\hat{U}(\boldsymbol{e}_{i_\tau}, \delta, \boldsymbol{B}_\tau, u_\tau) \leq \frac{1}{t_\tau} \leq L(\boldsymbol{A}_{i_\tau}, \delta_L, \boldsymbol{Q}_\tau, \ell_\tau)$.
5: Set $\boldsymbol{Q}_{\tau+1} = \boldsymbol{Q}_\tau + t_\tau \boldsymbol{A}_{i_\tau} \boldsymbol{A}_{i_\tau}^\top$, $\boldsymbol{B}_{\tau+1} = \boldsymbol{B}_\tau + t_\tau \boldsymbol{e}_{i_\tau} \boldsymbol{e}_{i_\tau}^\top$ and $S_{\tau+1, i_\tau} = \sqrt{r^{-1}(1 - \sqrt{n/r})}/\sqrt{t_\tau}$.
6: end for
7: return $\boldsymbol{S}$.

The theoretical guarantee of Algorithm 5.3 is given in the following Theorem 5.4 [4].

Theorem 5.4 *Algorithm 5.3 returns* $\hat{\boldsymbol{x}} \in \mathbb{R}^n$ *and* $\hat{\boldsymbol{\Pi}}$ *satisfying*

$$\|\hat{\boldsymbol{\Pi}}\boldsymbol{y} - \boldsymbol{A}\hat{\boldsymbol{x}}\|_2^2 \leq (1 + \epsilon) \min_{\boldsymbol{x}, \boldsymbol{\Pi}} \|\boldsymbol{\Pi}\boldsymbol{y} - \boldsymbol{A}\boldsymbol{x}\|_2^2.$$

It shows that Algorithm 5.3 enjoys recovery guarantees for $\boldsymbol{x}$ and $\boldsymbol{\Pi}$ when the data come from the Gaussian measurement model (5.8). Moreover, the overall running time is $\mathcal{O}((m/\epsilon)^{(k)})$ which is remarkably lower than that of naïve brute force search, i.e., $\Omega(n!)$.

Algorithm 5.3: Approximation algorithm for least squares problem (5.9)

Input Sample matrix $\boldsymbol{A} \in \mathbb{R}^{m \times n}$; observation $\boldsymbol{y} \in \mathbb{R}^m$; approximation parameter $\epsilon \in (0, 1)$. Assume $\boldsymbol{A}^\top \boldsymbol{A} = \boldsymbol{I}_n$.
Output Parameter vector $\hat{\boldsymbol{x}} \in \mathbb{R}^n$ and permutation matrix $\hat{\boldsymbol{\Pi}}$.
1: Run Algorithm 5.2 with input matrix $\boldsymbol{A}$ to obtain a matrix $\boldsymbol{S} \in \mathbb{R}^{r \times m}$ with $r = 4n$.
2: Let $\mathscr{B}$ be the set of vectors $\boldsymbol{b} = (b_1, b_2, \ldots, b_n)^\top \in \mathbb{R}^n$ satisfying the following: for each $i \in [n]$,

- if the i-th column of $\boldsymbol{S}$ is all zeros, then $b_i = 0$;
- otherwise, $b_i \in \{y_1, y_2, \ldots, y_n\}$.

3: Let $c := 1 + 4(1 + \sqrt{m/(4n)})^2$.
4: for each $\boldsymbol{b} \in \mathscr{B}$ do
5: Compute $\tilde{\boldsymbol{x}}_{\boldsymbol{b}} \in \operatorname{argmin}_{\boldsymbol{x} \in \mathbb{R}^n} \|\boldsymbol{S}(\boldsymbol{b} - \boldsymbol{A}\boldsymbol{x})\|_2^2$, and let $r_{\boldsymbol{b}} := \min_{\boldsymbol{\Pi}} \|\boldsymbol{\Pi}\boldsymbol{y} - \boldsymbol{A}\tilde{\boldsymbol{x}}_{\boldsymbol{b}}\|_2^2$.
6: Construct a $\sqrt{\epsilon r_{\boldsymbol{b}}/c}$-net $\mathscr{N}_{\boldsymbol{b}}$ for the Euclidean ball of radius $\sqrt{c r_{\boldsymbol{b}}}$ around $\tilde{\boldsymbol{x}}_{\boldsymbol{b}}$, so that for each $\boldsymbol{v} \in \mathbb{R}^k$ with $\|\boldsymbol{v} - \tilde{\boldsymbol{x}}_{\boldsymbol{b}}\|_2 \leq \sqrt{c r_{\boldsymbol{b}}}$, there exists $\boldsymbol{v}' \in \mathscr{N}_{\boldsymbol{b}}$ such that $\|\boldsymbol{v} - \boldsymbol{v}'\|_2 \leq \sqrt{\epsilon r_{\boldsymbol{b}}/c}$.
7: end for
8: return $\hat{\boldsymbol{x}} \in \underset{\boldsymbol{x} \in \bigcup_{\boldsymbol{b} \in \mathscr{B}} \mathscr{N}_{\boldsymbol{b}}}{\operatorname{argmin}} \min_{\boldsymbol{\Pi}} \|\boldsymbol{\Pi}\boldsymbol{y} - \boldsymbol{A}\boldsymbol{x}\|_2^2$ and $\hat{\boldsymbol{\Pi}} \in \underset{\boldsymbol{\Pi}}{\operatorname{argmin}} \|\boldsymbol{\Pi}\boldsymbol{y} - \boldsymbol{A}\hat{\boldsymbol{x}}\|_2^2$.

However, the approximation guarantee is not robust to even mild levels of noise. Thus, it motivates other advanced algorithms, e.g., alternative minimization approaches [1] and algebraic geometric approaches [15].

5.4 Algebraic-Geometric Approach

Recently, the paper [1] proposed a practical algorithm for solving shuffled linear regression (5.9) via alternating minimization: estimating $\boldsymbol{\Pi}^*$ via sorting an estimate $\boldsymbol{\xi}^*$ and estimating $\boldsymbol{\xi}^*$ via least-squares given an estimate $\boldsymbol{\Pi}^*$. Nevertheless, this approach is very sensitive to initialization and generally works only when the observation data is partially shuffled. To address the limitations of alternating minimization approach, the paper [15] proposed an algebraic geometric approach, which uses symmetric polynomials and leads to a polynomial system of n equations in n unknowns, containing $\boldsymbol{x}$ in its root locus.

Based on algebraic geometry theory, the paper [15] proved that this polynomial system is consistent with at most $n!$ complex roots as long as the independent samples are generic. This fact implies that this polynomial system can always be solved, and its most suitable root can be used as initialization to the expectation maximization (EM) algorithm.

5.4.1 Eliminating Π via Symmetric Polynomials

Prior to introducing the algebraically initialized expectation-maximization, we first describe the main idea of the algebraic-geometric approach to solve the shuffled linear regression estimation problem (5.8). Denote the ring of polynomials with real coefficients over variables

$$z := [z_1, \dots, z_m]^\top$$

as

$$\mathbb{R}[z] := \mathbb{R}[z_1, \dots, z_m].$$

A symmetric polynomial[1] $p \in \mathbb{R}[z]$ means that it is invariant to any permutation of the variables z, given by

$$p(z) := p(z_1, \dots, z_m) = p(z_{\pi(1)}, \dots, z_{\pi(m)}) =: p(\boldsymbol{\Pi} z), \tag{5.14}$$

where π is a permutation on $\{1, \dots, m\}$ and $\boldsymbol{\Pi}$ is an $m \times m$ permutation matrix.

Recall the shuffled linear regression problem (5.8) in the noiseless scenario and let $(\boldsymbol{\Pi}^*, \boldsymbol{x}^*)$ with

$$\boldsymbol{x}^* = [x_1^*, \dots, x_n^*]^\top$$

being a solution. Based on a symmetric polynomial $p \in \mathbb{R}[z]$, we get

$$\boldsymbol{\Pi}^* \boldsymbol{y} = \boldsymbol{A}\boldsymbol{x}^* \overset{p:\ \text{symmetric}}{\Longrightarrow} p(\boldsymbol{y}) = p(\boldsymbol{\Pi}^* \boldsymbol{y}) = p(\boldsymbol{A}\boldsymbol{x}^*). \tag{5.15}$$

In (5.15), the symmetric polynomial p plays a vital role in eliminating the unknown permutation $\boldsymbol{\Pi}^*$ and providing a constraint that only be relative with the known $\boldsymbol{A}, \boldsymbol{y}$,

$$\hat{p}(\boldsymbol{x}) := p(\boldsymbol{A}\boldsymbol{x}) - p(\boldsymbol{y}) = 0. \tag{5.16}$$

We aim to find the solution $\boldsymbol{x}^*$ that satisfies (5.16), thereby finding all solutions to the estimation problem (5.8).

To achieve this goal, we first introduce the concept of *algebraic variety* which is used to characterize the solutions of (5.16). Recall that the polynomial $\hat{p}$ in (5.16) is an element of the polynomial ring $\mathbb{R}[\boldsymbol{x}]$ in n variables $\boldsymbol{x} := [x_1, \dots, x_n]^\top$, and the set of its roots, denoted as

$$\mathcal{V}(\hat{p}) := \{\boldsymbol{x} \in \mathbb{R}^n : \hat{p}(\boldsymbol{x}) = 0\},$$

[1] We do not distinguish between p and $p(z)$.

is called an algebraic variety. In particular, $\mathscr{V}(\hat{p})$ defines a hypersurface of $\mathbb{R}^n$. Geometrically, the solutions to (5.16) are the intersection points of the corresponding n hypersurfaces

$$\mathscr{V}(\hat{p}_1), \ldots, \mathscr{V}(\hat{p}_n),$$

which include all solutions to problem (5.8), as well as potentially irrelevant points. Theorems provided in Sect. 5.4.2 investigate a system of n equations in n unknowns and the method of filtering its roots of interest is introduced in Sect. 5.4.3.

In addition, an example is provided in the following to illustrate the symmetric polynomial.

Example 5.2 Consider the data

$$\boldsymbol{A} = \begin{bmatrix} 1 & 2 \\ -2 & 4 \\ 0 & -5 \end{bmatrix}, \ \boldsymbol{y} = \begin{bmatrix} -20 \\ 11 \\ 10 \end{bmatrix}. \tag{5.17}$$

It is simple to find that there is a unique permutation

$$\boldsymbol{\Pi}^* = \begin{bmatrix} 0 & 1 & 0 \\ 0 & 0 & 1 \\ 1 & 0 & 0 \end{bmatrix} \tag{5.18}$$

that results in a consistent linear system of equations

$$\begin{bmatrix} 0 & 1 & 0 \\ 0 & 0 & 1 \\ 1 & 0 & 0 \end{bmatrix} \begin{bmatrix} -20 \\ 11 \\ 10 \end{bmatrix} = \begin{bmatrix} 1 & 2 \\ -2 & 4 \\ 0 & -5 \end{bmatrix} \begin{bmatrix} x_1 \\ x_2 \end{bmatrix} \tag{5.19}$$

with solution $\xi_1^* = 3$, $\xi_2^* = 4$. Now consider the symmetric polynomial

$$p_1(z_1, z_2, z_3) = z_1 + z_2 + z_3, \tag{5.20}$$

and based on (5.16) it yields the constraint

$$(x_1 + 2x_2) + (4x_2 - 2x_1) - 5x_2 = -20 + 11 + 10, \tag{5.21}$$

$$\Leftrightarrow x_2 - x_1 = 1, \tag{5.22}$$

$$\Leftrightarrow \hat{p}_1(\boldsymbol{x}) := p_1(\boldsymbol{A}\boldsymbol{x}) - p_1(\boldsymbol{y}) = 0. \tag{5.23}$$

Indeed, we see that the solution $\boldsymbol{\xi}^* = [3,\ 4]^\top$ satisfies (5.23).

5.4.2 Theoretical Analysis

The theoretical analysis on symmetric polynomials for both exact and corrupted data are provided in the following.

5.4.2.1 Exact Data

As Example 5.2 suggests, a natural choice for n symmetric polynomials are the first n power sums

$$p_k(z) \in \mathbb{R}[z] := \mathbb{R}[z_1, \ldots, z_m],\ k \in [n] := \{1, \ldots, n\},$$

denoted as

$$p_k(z) := z_1^k + \cdots + z_m^k. \tag{5.24}$$

Based on (5.16), we conclude that any solution $\boldsymbol{\xi}^*$ of (5.8) in the noiseless scenario must obey the polynomial constraints

$$\hat{p}_k(\boldsymbol{x}) = 0, \quad k \in [n], \text{ where} \tag{5.25}$$

$$\hat{p}_k(\boldsymbol{x}) := p_k(\boldsymbol{A}\boldsymbol{x}) - p_k(\boldsymbol{y}) = \sum_{i=1}^{m} (\boldsymbol{a}_i^\top \boldsymbol{x})^k - \sum_{j=1}^{m} y_j^k, \tag{5.26}$$

and $\boldsymbol{a}_i^\top$ presents the ith row of $\boldsymbol{A}$. The following theorem provides a theoretical guarantee that the number of other irrelevant solutions must be finite.

Theorem 5.5 ([15]) *Assuming* $\boldsymbol{A}$ *is generic and* $\boldsymbol{y}$ *is some permutation of a vector. The algebraic variety*

$$\mathscr{V}(\hat{p}_1, \ldots, \hat{p}_n)$$

consists of all

$$\boldsymbol{\xi}_1^*, \ldots, \boldsymbol{\xi}_\ell^* \in \mathbb{R}^n$$

such that there exists permutations

$$\boldsymbol{\Pi}_1^*, \ldots, \boldsymbol{\Pi}_\ell^*$$

with

$$\boldsymbol{\Pi}_i^* \boldsymbol{y} = \boldsymbol{A}\boldsymbol{\xi}_i^*,\ \forall i \in [\ell],$$

while it may include at most $n! - \ell$ *other points of* $\mathbb{C}^n$.

Theorem 5.5 demonstrates that the system of polynomial equations

$$\hat{p}_1(\boldsymbol{x}) = \cdots = \hat{p}_n(\boldsymbol{x}) = 0, \tag{5.27}$$

always has a finite number of solutions in $\mathbb{C}^n$ (at most $n!$), which contain all possible solutions $\boldsymbol{\xi}_1^*, \ldots, \boldsymbol{\xi}_\ell^* \in \mathbb{R}^n$ of problem (5.8).

5.4.2.2 Corrupted Data

The following theorem addresses the issue of corrupted data which is common in practical applications. Considering the corrupted data which is denoted as $\tilde{\boldsymbol{A}}, \tilde{\boldsymbol{y}}$, the linear system can be represented as

$$\boldsymbol{\Pi} \tilde{\boldsymbol{y}} = \tilde{\boldsymbol{A}} \boldsymbol{x}. \tag{5.28}$$

There exists a permutation $\boldsymbol{\Pi} = \tilde{\boldsymbol{\Pi}}^*$ such that (5.28) is approximately consistent, if the degree of corruption is sufficiently small. In order to get an approximate solution of (5.28), a *corrupted* power-sum polynomial is defined as

$$\tilde{p}_k(\boldsymbol{x}) := p_k(\tilde{\boldsymbol{A}} \boldsymbol{x}) - p_k(\tilde{\boldsymbol{y}}), \quad k \in [n], \tag{5.29}$$

and the polynomial system $\tilde{\mathscr{P}}$ is considered, given by

$$\tilde{p}_1 = \cdots = \tilde{p}_n = 0. \tag{5.30}$$

These are n equations of degrees $1, 2, \ldots, n$ with n unknowns. The system of polynomial equations (5.30) with respect to corrupted data is investigated in the following theorem.

Theorem 5.6 ([15]) *If $\tilde{\boldsymbol{A}}$ is generic and $\tilde{\boldsymbol{y}} \in \mathbb{R}^m$ is any vector, then $\mathscr{V}(\tilde{p}_1, \ldots, \tilde{p}_n)$ is non-empty containing at most $n!$ points of $\mathbb{C}^n$.*

Theorem 5.6 demonstrates that the system of polynomial equations (5.30) always has at least one solution. We can conclude that an approximate solution to the shuffled linear system (5.28) lies in a finite number of solutions of the system (5.30). Theorems 5.5 and 5.6 provide theoretical guarantees for developing algebraical method to solve the shuffled linear regression problem. The algorithm, called algebraically initialized expectation-maximization, is introduced in the next section.

5.4.3 Algebraically Initialized Expectation-Maximization

If there is a unique solution $\boldsymbol{\xi}^*$ to the shuffled linear regression problem (5.8), Theorem 5.5 ensures that $\boldsymbol{\xi}^*$ is one of the finitely many complex roots of the polynomial system (5.25) of n equations in n unknowns. Moreover, in the case of corrupted data, Theorem 5.6 ensures that the system is consistent with $L \leq n!$ complex roots, and if the corruption degree is modest, one of the roots can be a good approximation to the maximum likelihood estimator (MLE) $\hat{\boldsymbol{\xi}}_{\mathrm{ML}}$ (5.9). Thus, the goal is to filter that root and refine it.

Particularly, several state-of-the-art polynomial system solvers [5] can be exploited to solve the polynomial system of equations (5.25). With the computed roots

$$\hat{\boldsymbol{\xi}}_1, \ldots, \hat{\boldsymbol{\xi}}_L \in \mathbb{C}^n,\ L \leq n!,$$

of the polynomial system, only their real parts

$$(\hat{\boldsymbol{\xi}}_1)_{\mathbb{R}}, \ldots, (\hat{\boldsymbol{\xi}}_L)_{\mathbb{R}}$$

are retained, which can be used for obtaining an approximation to the ML estimator $\hat{\boldsymbol{\xi}}_{\mathrm{ML}}$. This is achieved by selecting the root that yields the smallest ℓ_2 error among all possible permutations $\boldsymbol{\Pi}$:

$$\hat{\boldsymbol{\xi}}_{\mathrm{AI}} := \underset{i \in [L]}{\operatorname{argmin}} \left\{ \min_{\boldsymbol{\Pi}} \left\| \boldsymbol{\Pi} \boldsymbol{y} - \boldsymbol{A} (\hat{\boldsymbol{\xi}}_i)_{\mathbb{R}} \right\|_2 \right\}. \tag{5.31}$$

Furthermore, the *algebraic initialization* $\hat{\boldsymbol{\xi}}_{\mathrm{AI}}$ is utilized as an initialization to the expectation maximization algorithm [1] which implements alternating minimization to solve (5.9). This method is called *Algebraically Initialized Expectation-Maximization (AI-EM)* [15] illustrated in Algorithm 5.4.

Algorithm 5.4: Algebraically initialized expectation-maximization

---

procedure AI-EM($\boldsymbol{y} \in \mathbb{R}^m,\ \boldsymbol{A} \in \mathbb{R}^{m \times n},\ T \in \mathbb{N}, \epsilon \in \mathbb{R}_+$)
1: $p_k(\boldsymbol{z}) := \sum_{j=1}^m z_j^k,\ \hat{p}_k := p_k(\boldsymbol{A}\boldsymbol{x}) - p_k(\boldsymbol{y}),\ k \in [n]$;
2: Compute roots $\{\hat{\boldsymbol{\xi}}_i\}_{i=1}^L \subset \mathbb{C}^n$ of $\{\hat{p}_k = 0,\ k \in [n]\}$;
3: Extract the real parts $\{(\hat{\boldsymbol{\xi}}_i)_{\mathbb{R}}\}_{i=1}^L$ of $\{\hat{\boldsymbol{\xi}}_i\}_{i=1}^L \subset \mathbb{C}^n$;
4: $\{\boldsymbol{\xi}_0, \boldsymbol{\Pi}_0\} \leftarrow \operatorname{argmin}_{\boldsymbol{\xi} \in \{(\hat{\boldsymbol{\xi}}_i)_{\mathbb{R}}\}_{i=1}^L, \boldsymbol{\Pi}} \|\boldsymbol{\Pi}\boldsymbol{y} - \boldsymbol{A}\boldsymbol{\xi}\|_2$;
5: $t \leftarrow 0, \Delta\mathcal{J} \leftarrow \infty,\ \mathcal{J} \leftarrow \|\boldsymbol{\Pi}_0\boldsymbol{y} - \boldsymbol{A}\boldsymbol{\xi}_0\|_2$;
6: while $t < T$ and $\Delta\mathcal{J} > \varepsilon\mathcal{J}$ do
7: $\quad t \leftarrow t + 1$;
8: $\quad \boldsymbol{\xi}_t \leftarrow \operatorname{argmin}_{\boldsymbol{\xi} \in \mathbb{R}^n} \|\boldsymbol{\Pi}_{t-1}\boldsymbol{y} - \boldsymbol{A}\boldsymbol{\xi}\|_2$;
9: $\quad \boldsymbol{\Pi}_t \leftarrow \operatorname{argmin}_{\boldsymbol{\Pi}} \|\boldsymbol{\Pi}\boldsymbol{y} - \boldsymbol{A}\boldsymbol{\xi}_t\|_2$;
10: $\quad \Delta\mathcal{J} \leftarrow \mathcal{J} - \|\boldsymbol{\Pi}_t\boldsymbol{y} - \boldsymbol{A}\boldsymbol{\xi}_t\|_2$;
11: $\quad \mathcal{J} \leftarrow \|\boldsymbol{\Pi}_t\boldsymbol{y} - \boldsymbol{A}\boldsymbol{\xi}_t\|_2$;
12: end while
13: Return $\boldsymbol{\xi}_t,\ \boldsymbol{\Pi}_t$.
end procedure

5.4.4 Simulation Results

To further illustrate the advantage of Algorithm 5.4, we compare it with two variations of EM algorithms that were proposed in [1]: (1) *LS-EM* that computes the MLE via alternating minimization with the initialization satisfying

$$\boldsymbol{\xi}_{0,\mathrm{LS}} := \underset{\boldsymbol{\xi}\in\Re^n}{\operatorname{argmin}} \|\boldsymbol{y} - \boldsymbol{A}\boldsymbol{\xi}\|_2; \tag{5.32}$$

(2) *Soft-EM* that uses the same initialization as LS-EM, but exploits a dynamic empirical average of permutation matrices drawn from a suitable Markov chain to optimize the permutation operation.

All methods are evaluated by measuring the relative estimation error between the estimator $\hat{\boldsymbol{\xi}}$ and the ground truth $\boldsymbol{\xi}^*$, given by

$$100\frac{\|\boldsymbol{\xi}^* - \hat{\boldsymbol{\xi}}\|_2}{\|\boldsymbol{\xi}^*\|_2}\%. \tag{5.33}$$

For AI-EM, the estimation error between the best root $\boldsymbol{\xi}^*_{\mathrm{AI}}$ of the polynomial system is defined as

$$\boldsymbol{\xi}^*_{\mathrm{AI}} := \underset{\hat{\boldsymbol{\xi}}_i,\, i\in[L]}{\operatorname{argmin}} \|\boldsymbol{\xi}^* - (\hat{\boldsymbol{\xi}}_i)_{\Re}\|_2, \tag{5.34}$$

and the estimator $\hat{\boldsymbol{\xi}}_{\mathrm{AI}}$ is computed as in (5.31).

Figure 5.2 illustrates the estimation error of the three methods with fully shuffled data under the setting of $n = 3$, $\sigma = 0$:0.01:0.1 and $m = 500$. The simulation

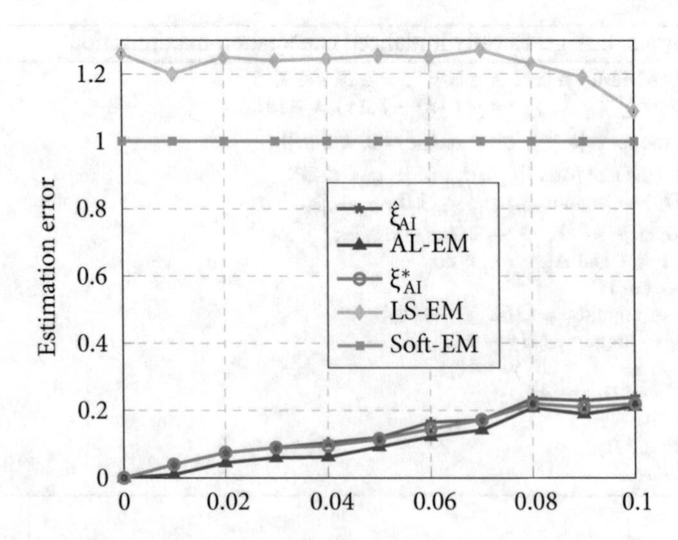

Fig. 5.2 Estimation error for fully shuffled data

results are averaged over 100 independent trials. It shows LS-EM and Soft-EM fail. It can be explained that when the data are fully shuffled, the least-squares initialization (5.32) exploited by both LS-EM and Soft-EM rather deviates from the ground truth $\boldsymbol{\xi}^*$.

5.5 Summary

This chapter summarized a shuffled linear regression model to support joint data decoding and device identification, thereby reducing the overhead in massive connectivity systems. The methods developed for solving the shuffled linear regression estimation problem are presented in this chapter from the numerical and theoretical points of view. The methods can be mainly categorized into two types: maximum likelihood estimation based approach and algebraic geometric approach. Besides the application introduced in this chapter, the shuffled linear regression method and its variations arise in many applications, e.g., image processing [6], user de-anonymization [8], and correspondence estimation [7]. Recently, an abstraction of shuffled linear problems which is called homomorphic sensing has been studied in [14], and an algebraic theory for homomorphic sensing has been developed. The paper [14] provides the first working solutions for the unlabeled sensing problem for small dimensions. It is still a principle direction of study to develop more efficient algorithms and corresponding theoretical guarantees for homomorphic sensing.

References

1. Abid, A., Zou, J.: A stochastic expectation-maximization approach to shuffled linear regression. In: Proceedings of the 56th Annual Allerton Conference on Communication, Control, and Computing, pp. 470–477. IEEE, Piscataway (2018)
2. Abid, A., Poon, A., Zou, J.: Linear regression with shuffled labels (2017). Preprint. arXiv:1705.01342
3. Boutsidis, C., Drineas, P., Magdon-Ismail, M.: Near-optimal coresets for least-squares regression. IEEE Trans. Inf. Theory. **59**(10), 6880–6892 (2013)
4. Hsu, D.J., Shi, K., Sun, X.: Linear regression without correspondence. In: Advances in Neural Information Processing Systems (NeurIPS), pp. 1531–1540 (2017)
5. Lazard, D.: Thirty years of polynomial system solving, and now? J. Symb. Comput. **44**(3), 222–231 (2009)
6. Lian, W., Zhang, L., Yang, M.H.: An efficient globally optimal algorithm for asymmetric point matching. IEEE Trans. Pattern Anal. Mach. Intell. **39**(7), 1281–1293 (2016)
7. Marques, M., Stosić, M., Costeira, J.: Subspace matching: unique solution to point matching with geometric constraints. In: Proceedings of the International Conference on Computer Vision, pp. 1288–1294. IEEE, Piscataway (2009)
8. Narayanan, A., Shmatikov, V.: Robust de-anonymization of large sparse datasets. In: Proceedings of the IEEE Symposium on Security and Privacy, pp. 111–125 (2008)

9. Pananjady, A., Wainwright, M.J., Courtade, T.A.: Linear regression with shuffled data: statistical and computational limits of permutation recovery. IEEE Trans. Inf. Theory **64**(5), 3286–3300 (2018)
10. Peng, L., Song, X., Tsakiris, M.C., Choi, H., Kneip, L., Shi, Y.: Algebraically-initialized expectation maximization for header-free communication. In: Proceedings of the IEEE International Conference on Acoustics Speech Signal Processing (ICASSP), pp. 5182–5186 (2019)
11. Pradhan, S.S., Kusuma, J., Ramchandran, K.: Distributed compression in a dense microsensor network. IEEE Signal Process. Manag. **19**(2), 51–60 (2002)
12. Scaglione, A., Servetto, S.: On the interdependence of routing and data compression in multi-hop sensor networks. Wirel. Netw. **11**(1–2), 149–160 (2005)
13. Slawski, M., Ben-David, E., et al.: Linear regression with sparsely permuted data. Electron. J. Stat. **13**(1), 1–36 (2019)
14. Tsakiris, M.C., Peng, L.: Homomorphic sensing. In: Proceedings of the International Conference on Machine Learning (ICML), pp. 6335–6344 (2019)
15. Tsakiris, M.C., Peng, L., Conca, A., Kneip, L., Shi, Y., Choi, H.: An algebraic-geometric approach to shuffled linear regression (2018). Preprint. arXiv: 1810.05440
16. Unnikrishnan, J., Haghighatshoar, S., Vetterli, M.: Unlabeled sensing with random linear measurements. IEEE Trans. Inf. Theory. **64**(5), 3237–3253 (2018)

Chapter 6
Learning Augmented Methods

Abstract In this chapter, we introduce some cutting-edge learning augmented techniques to enhance the performance of structured signal processing. We start with compressed sensing under a generative prior, which can better capture the underlying signal structure than the traditional sparse prior. We then present learning augmented techniques for the joint design of measurement matrix and sparse support recovery for the sparse linear model (e.g., compressed sensing). Furthermore, several deep-learning-based AMP methods for the sparse linear model are introduced, including learned AMP, learned Vector-AMP, and learned ISTA for group row sparsity.

6.1 Structured Signal Processing Under a Generative Prior

Recall the sparse linear model defined in (2.3). The sparse signal $\boldsymbol{x}$ can be recovered via solving a convex optimization problem known as Lasso [21]:

$$\hat{\boldsymbol{x}} = \arg\min_{\boldsymbol{x}} \frac{1}{2}\|\boldsymbol{y} - \boldsymbol{A}\boldsymbol{x}\|_2^2 + \lambda\|\boldsymbol{x}\|_1, \tag{6.1}$$

where the parameter $\lambda > 0$ controls the sparsity level. Instead of focusing on the sparsity of $\boldsymbol{x}$ in (2.3), the paper [3] has recently estimated $\hat{\boldsymbol{x}}$ based on the structure derived from a *generative model*. It demonstrates that the data distribution can be identified by neural network based generative models, e.g., variational auto-encoders (VAEs) [12] and generative adversarial networks (GANs) [8]. The neural network based generative model learns a generator $G(z) : z \in \mathbb{R}^k \rightarrow G(z) \in \mathbb{R}^n$ that maps a low dimensional space z to the high-dimensional sample space. This generator is trained to generate vectors that approximate the vectors in the training dataset. Here, the generator characterizes a probability distribution over vectors in the sample space, and based on the training dataset, the generator is trained to allocate higher probabilities to more likely vectors. Thus, the notion of a vector in certain space can be generally captured by a pre-trained generator. It was shown in

Y. Shi et al., *Low-overhead Communications in IoT Networks*,
https://doi.org/10.1007/978-981-15-3870-4_6

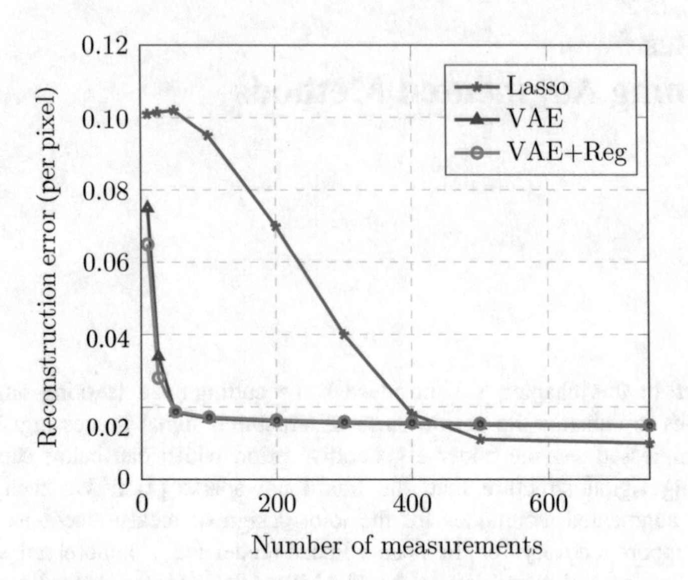

Fig. 6.1 The performance of compressed sensing under generative models

[3] that a vector in the space of a given system setting is close to some point in the range of G.

The paper [3] also proposed an algorithm that exploits generative models to solve the compressed sensing problem. This algorithm optimizes the variable $z \in \mathbb{R}^k$ via gradient descent such that the corresponding generator $G(z)$ yields a small measurement error, i.e.,

$$\|\boldsymbol{A}G(z) - \boldsymbol{y}\|_2^2. \tag{6.2}$$

Even though the objective function (6.2) is nonconvex, it was empirically demonstrated in [3] that gradient descent works well, and can yield significantly better performance than Lasso with relatively few measurements. Figure 6.1 illustrates the comparison of signal reconstruction from compressed linear measurements with the sparse linear model and a generative model, i.e., VAE. The experiment is run on the MNIST dataset, a classic 10-class hand-written digit classification dataset. Each pixel value of an image is either 0 (background) or 1 (foreground), so the digit images are reasonably sparse in the pixel space. Three different algorithms are considered: the classic Lasso algorithm for the sparse linear model in the pixel space, and two algorithms, "VAE" and "VAE+Reg," for the generative model with loss functions

$$\|\boldsymbol{A}G(z) - \boldsymbol{y}\|_2^2,$$

and

$$\|AG(z) - y\|_2^2 + \lambda \|z\|_2,$$

where $G(z)$ is the generative model trained on the MNIST dataset. This result shows that signal reconstruction with generative models requires 10× fewer measurements than a conventional sparse linear model to achieve 10% error.

The paper [3] provided the theoretical performance guarantee. Specifically, it was demonstrated that, as long as a good approximate solution to the objective (6.2) is found by gradient descent, the generator $G(z)$, which yields the closest possible point in the range of G, will be sufficiently close to the ground truth x^*. The proof provided in [3] relies on the set-restricted eigenvalue condition which is a generalization of the restricted eigenvalue condition (REC). Moreover, [3] shows that for some generators, e.g., VAEs and GANs, random Gaussian measurement matrices can satisfy the set-restricted eigenvalue condition with high probability. That is, for d-layer neural networks, $O(kd \log n)$ Gaussian measurements sufficiently guarantee high-accuracy reconstruction with high probability. Specifically, the result states as below.

Theorem 6.1 ([3]) *Assuming that* $G : \mathbb{R}^k \to \mathbb{R}^n$ *is an L-Lipschitz function, and* $A \in \mathbb{R}^{m\times n}$ *is a random Gaussian matrix for* $m = O(k \log \frac{Lr}{\delta})$*, obeying* $A_{ij} \sim N(0, 1/m)$*. For any* $x^* \in \mathbb{R}^n$ *and observation* $y = Ax^* + \eta$*, let the estimator* $\hat{z}$ *minimize (6.2) to within additive* ε *of the optimum over vectors with* $\|\hat{z}\|_2 \leq r$*. Then with* $1 - e^{-\Omega(m)}$ *probability, there is*

$$\|G(\hat{z}) - x^*\|_2 \leq 6 \min_{\substack{z^* \in \mathbb{R}^k \\ \|z^*\|_2 \leq r}} \|G(z^*) - x^*\|_2 + 3\|\eta\|_2 + 2\varepsilon + 2\delta. \tag{6.3}$$

The first two terms on the right-hand side of (6.3) identify the minimum error of any vector in the range of the generator and the norm of the noise, respectively. The third term ε comes from the distance between the global optimum and the convergence result generated by gradient descent.

The results of [3] have inspired lots of follow-up studies. Recently, the paper [23] proposed a novel framework that significantly improves both the performance and speed of signal recovery by jointly training a generator and the optimization process for reconstruction via meta-learning. The paper explored training the measurements with different objectives, and derived a family of models based on minimizing measurement errors. We will provide an overview of this work in Sect. 6.2.

Besides the compressed sensing problem, the generative prior has also been applied to the blind image deconvolution problem [1]. It can also provide some insights on applying the generative prior in the blind demixing problem introduced in Chap. 3 and the sparse blind demixing problem introduced in Chap. 4.

6.2 Joint Design of Measurement Matrix and Sparse Support Recovery

The design of the measurement matrix in compressed sensing is of critical importance for both practical implementation and performance enhancement (e.g., achieving a better compression or allowing a higher signal reconstruction quality). Thus, it has received intensive attentions, and good progresses have been made. Learning augmented techniques have recently exploited in joint design of measurement matrix and sparse support recovery. This section first introduces basic methods for solving this problem, followed by the learning augmented methods.

Sample Scheduling The paper [10] proposed an adaptive CS based sample scheduling mechanism (ACS) with respect to different per-sampling-window bases for wireless sensor networks. For each basis, given a sensing quality, ACS estimates the minimum required sample rate, thereby correspondingly adjusting sensors' sample rates.

Sensing Matrix Optimization To optimize sensing matrices, some techniques known as mutual coherence minimization [6, 7, 14, 24] are developed, without additional assumption on the class of acquired signals. Another line of research shows that better results can be achieved with some priors on the input signal. For instance, when the energy of the signals to be acquired is not evenly distributed, i.e., when they are both sparse and localized. Mathematically, for the sparse signal $\boldsymbol{x}$ in compressed sensing (1.5), it holds that $\mathbb{E}(\boldsymbol{x}\boldsymbol{x}^{\top})$ is not a multiplier of the $n \times n$ identity matrix $\boldsymbol{I}_n$. To characterize this property, the paper [14] introduced a design criterion, which is called *rakeness*, to identify the amount of energy that the measurements seize from the acquired signal. The proposed *rakeness* approach [14] aligns statistical properties of the compressed sensing stage with that of the input signal $\boldsymbol{x}$, while simultaneously preserving conditions for a correct signal reconstruction required by the standard compressed sensing theory.

Following the idea proposed in [14], the paper [15] proposed sensing matrix optimization techniques that exploit statistical properties of the process generating $\boldsymbol{x}$ in compressed sensing (1.5). One method is nearly orthogonal CS, which is based on a geometric constraint enforcing diversity between different compressed measurements. Another method, named, maximum-energy CS, is a heuristic screening of candidate measurements that relies on a self-adapted optimization procedure.

Learning Augmented Methods Recent works [13, 17, 18, 22, 23] consider joint design of signal compression and recovery methods using auto-encoder [13, 17, 18, 22] and generative adversarial networks (GAN) [23] in deep learning. In particular, linear compression for real signals was considered in [18, 22]; nonlinear compression for real signals was considered in [17, 23]. The paper [13] studied linear compression for complex signals.

The fundamental idea of joint design of signal compression and recovery methods using auto-encoder [13, 17, 18] can be summarized as follows.

- Collect an input signal $\boldsymbol{x}^{(i)}$ from a training set $\mathscr{D}_{\text{train}} = \{\boldsymbol{x}^{(i)}\}_{i=1}^{s}$.
- Reconstruct input's components and reduce its dimensionality via d layer operations such as convolutional layer [17], linear reduction mapping [18], etc. Denote the set of parameters at d layers as $\Omega = \{\boldsymbol{W}_j, \boldsymbol{b}_j\}_{j=1}^{d}$.
- Take undersampled measurements.
- Increase measurements dimensionality via operations such as convolutional layers [17], nonlinear inverse mapping [18], etc.
- Convert the output to a reconstructed signal. Denote this mapping from original signals to reconstructed signals as $\hat{\boldsymbol{x}} = \mathscr{F}(\boldsymbol{x}, \Omega)$.

The mean-squared error (MSE) can be adopted as a loss function over the training set $\mathscr{D}_{\text{train}}$

$$\mathscr{L}(\Omega) = \frac{1}{s}\sum_{i=1}^{s}\left\|\mathscr{F}(\boldsymbol{x}, \Omega) - \boldsymbol{x}^{(i)}\right\|_2^2. \tag{6.4}$$

The stochastic gradient descent (SGD) or ADMM optimizer [11] can be applied for minimizing $\mathscr{L}(\Omega)$ (6.4) and learning parameters. Recently, the work [13] extended the methods in [17, 18] to joint linear compression and recovery methods for complex signal estimation, which is more challenging. The proposed architecture includes two components, an auto-encoder and a hard thresholding module. The proposed auto-encoder successfully deals with complex signals via exploiting standard auto-encoder for real numbers. The key technique is to establish the encoder which mimics the noisy linear measurement process. The model for complex numbers in compressed sensing, i.e.,

$$\boldsymbol{y} = \boldsymbol{A}\boldsymbol{x} + \boldsymbol{z},$$

can be equivalently expressed via the following two expressions of real numbers:

$$\Re(\boldsymbol{y}) = \Re(\boldsymbol{A})\Re(\boldsymbol{x}) - \Im(\boldsymbol{A})\Im(\boldsymbol{x}) + \Re(\boldsymbol{z}), \tag{6.5}$$

$$\Im(\boldsymbol{y}) = \Im(\boldsymbol{A})\Re(\boldsymbol{x}) + \Re(\boldsymbol{A})\Im(\boldsymbol{x}) + \Im(\boldsymbol{z}). \tag{6.6}$$

Besides the decoders mentioned above, a recent paper [22] presented a ℓ_1 decoder to learn linear encoders that adjust to data. The convex and non-smooth ℓ_1 decoder cannot be trained via standard gradient-based, i.e., gradient propagation. To address this issue, the paper [22] relies on the idea of unrolling the convex decoder into T projected subgradient steps. Denote the ℓ_1-minimization as:

$$L(\boldsymbol{A}, \boldsymbol{y}) := \arg\min_{\boldsymbol{x}\in\mathbb{R}^d} \|\boldsymbol{x}\|_1 \quad \text{s.t. } \boldsymbol{A}\boldsymbol{x} = \boldsymbol{y}. \tag{6.7}$$

Mathematically, given a training set $\mathcal{D}_{\text{train}}$, the problem of finding the best $\boldsymbol{A}$ can be formulated as

$$\min_{\boldsymbol{A}\in\mathbb{R}^{m\times d}} f(\boldsymbol{A}) := \sum_{i=1}^{s} \|\boldsymbol{x}^{(i)} - L(\boldsymbol{A}, \boldsymbol{A}\boldsymbol{x}^{(i)})\|_2^2.$$

Here $L(\cdot,\cdot)$ is the ℓ_1 decoder defined in (6.7). Unfortunately, it is difficult to compute the gradient of $f(\boldsymbol{A})$ with respect to $\boldsymbol{A}$, due to the optimization problem defined in (6.7). The paper [22] addressed this issue by replacing the ℓ_1-minimization with the iterations of T-step projected subgradient, which approximately computes the gradients. Define an approximate function $\tilde{f}(\boldsymbol{A}) : \mathbb{R}^{m\times d} \mapsto \mathbb{R}$, and this procedure can be represented as

$$\begin{aligned} \tilde{f}(\boldsymbol{A}) &:= \sum_{i=1}^{s} \|\boldsymbol{x}^{(i)} - \hat{\boldsymbol{x}}^{(i)}\|_2^2, \quad \text{where} \\ \hat{\boldsymbol{x}}^{(i)} &= T\text{-step projected subgradient of} \\ &\quad\; L(\boldsymbol{A}, \boldsymbol{A}\boldsymbol{x}^{(i)}), \text{ for } i = 1, \ldots, s, \end{aligned} \tag{6.8}$$

which is called *unrolling*.

Another type of data-driven approaches is using learning augmented generative adversarial networks (GAN) [23] for joint design of signal compression and recovery methods. The paper [23] generalized the measurement matrix $\boldsymbol{A}$ in compressed sensing under a generative prior $G_\phi(\cdot)$, i.e., (6.2). To achieve this, the paper [23] defines a measurement function $\boldsymbol{y} \leftarrow F_\phi(\boldsymbol{x})$ where $\boldsymbol{x} = G_\phi(\boldsymbol{z})$, thus both F_ϕ and G_ϕ can be trained via deep neural networks. The key point of this generalized setting is recovering the signal $\boldsymbol{x}$ from inverting the measurement function $\boldsymbol{x} \leftarrow F_\phi^{-1}(\boldsymbol{y})$ via minimizing the measurement error:

$$E_\phi(\boldsymbol{y}, \boldsymbol{z}) = \|\boldsymbol{y} - F_\phi\left(G_\phi(\boldsymbol{z})\right)\|_2^2. \tag{6.9}$$

6.3 Deep-Learning-Based AMP

Deep learning recently has achieved great successes in many applications, which has inspired recent developments of deep-learning-based methods for structured signal processing. In this section, we introduce two neural network architectures proposed in [4]. Similar to the approximate message passing (AMP) algorithms that decouple prediction errors across iterations, the deep-learning-based AMP in [4] decouples prediction errors across layers. In [4], the proposed methods were applied to solve the compressive random access and massive-MIMO channel estimation in 5G networks. These methods also bring some insights on developing deep-learning-based AMP to solve the joint device activity detection and channel estimation problem in massive IoT networks.

We first introduce the "learned AMP" network proposed in [4], followed by the "learned VAMP" network [4] that offers increased robustness to deviations in the i.i.d. Gaussian measurement matrix. In both cases, the linear transforms and scalar nonlinearities of the network are simultaneously learned. An straightforward interpretation of learned VAMP is demonstrated in [4] that with i.i.d. measurements, the linear transforms and scalar nonlinearities established by the VAMP algorithm follow the values learned through back-propagation. Furthermore, we introduce a learned-based algorithm for group row sparsity (LISTA-GS) to estimate the sparse linear model (2.8) in the multiple-antenna scenario in Sect. 6.3.3.

$$\boldsymbol{Y} = \boldsymbol{Q}\boldsymbol{\Theta} + \boldsymbol{N}. \tag{6.10}$$

Besides the approaches introduced in this section, the authors in [9, 16, 20, 25] exploited properties of sparsity patterns of real signals [9, 16, 25] and complex signals [20] from training samples using data-driven approaches based on deep learning, which also brings some insights for future study.

6.3.1 Learned AMP

For compressed sensing (2.3), i.e.,

$$\boldsymbol{y} = \boldsymbol{A}\boldsymbol{x} + \boldsymbol{n},$$

where $\boldsymbol{y} \in \mathbb{C}^L$, $\boldsymbol{x} \in \mathbb{C}^N$,

$$\boldsymbol{n} \in \mathbb{C}^L \sim \mathcal{CN}(\boldsymbol{0}, \sigma^2\boldsymbol{I}) \tag{6.11}$$

is the additive white Gaussian noise, the sparse signal $\boldsymbol{x}$ can be estimated by Lasso (6.1) given by

$$\hat{\boldsymbol{x}} = \arg\min_{\boldsymbol{x}} \frac{1}{2}\|\boldsymbol{y} - \boldsymbol{A}\boldsymbol{x}\|_2^2 + \lambda\|\boldsymbol{x}\|_1.$$

It can be solved by the approximate message passing algorithm (2.27) introduced in Sect. 2.4.1:

$$\boldsymbol{r}_t = \boldsymbol{y} - \boldsymbol{A}\hat{\boldsymbol{x}}_t + \frac{1}{M}\|\hat{\boldsymbol{x}}_t\|_0\boldsymbol{r}_{t-1} \tag{6.12a}$$

$$\hat{\boldsymbol{x}}_{t+1} = \eta_{st}\left(\hat{\boldsymbol{x}}_t + \boldsymbol{A}^{\mathsf{T}}\boldsymbol{r}_t; \frac{\alpha}{\sqrt{M}}\|\boldsymbol{r}_t\|_2\right), \tag{6.12b}$$

where the initial points are set as $\hat{\boldsymbol{x}}_0 = \mathbf{0}$, $\boldsymbol{r}_{-1} = \mathbf{0}$, $t \in \{0, 1, 2, \dots\}$, and $\eta_{st}(\cdot; \lambda) : \mathbb{R}^N \to \mathbb{R}^N$ is the "soft thresholding" shrinkage function, component wisely defined as

$$[\eta_{st}(\boldsymbol{r}; \lambda)]_i \triangleq \text{sgn}(r_i) \max\{|r_i| - \lambda, 0\}. \tag{6.13}$$

In (6.12b), α is a tuning parameter that correlates with λ in (6.1). The AMP-inspired deep networks for solving sparse linear problems have been proposed in [4], which are introduced in the sequel.

The paper [4] established a neural network via unfolding the iterations of AMP-ℓ_1 from (6.12), followed by learning the MSE-minimal values of the network parameters, which is called "LAMP-ℓ_1." The t-th layer of the LAMP-ℓ_1 network is represented by

$$\hat{\boldsymbol{x}}_{t+1} = \beta_t \eta_{st}\left(\hat{\boldsymbol{x}}_t + \boldsymbol{B}_t \boldsymbol{r}_t; \frac{\alpha_t}{\sqrt{M}} \|\boldsymbol{r}_t\|_2\right) \tag{6.14a}$$

$$\boldsymbol{r}_{t+1} = \boldsymbol{y} - \boldsymbol{A}\hat{\boldsymbol{x}}_{t+1} + \frac{\beta_t}{M} \|\hat{\boldsymbol{x}}_{t+1}\|_0 \boldsymbol{r}_t, \tag{6.14b}$$

where the first-layer inputs are set as $\hat{\boldsymbol{x}}_0 = \mathbf{0}$ and $\boldsymbol{r}_0 = \boldsymbol{y}$. The paper [4] refers to networks that use fixed $\boldsymbol{B}$ over all layers t as "tied," where the LAMP-ℓ_1 parameters are $\boldsymbol{\Omega} = \left\{\boldsymbol{B}, \{\alpha_t, \beta_t\}_{t=0}^{T-1}\right\}$. Those that depend on t, i.e., $\boldsymbol{B}_t$, are named as "untied," where the LAMP-ℓ_1 parameters are $\boldsymbol{\Omega} = \{\boldsymbol{B}_t, \alpha_t, \beta_t\}_{t=0}^{T-1}$. The parameters $\boldsymbol{\Omega}$ of "tied" case and "untied" case can be further learned by minimizing the MSE on the training data, which are illustrated in Algorithms 6.1 and 6.2, respectively. In [4], it was demonstrated that LAMP-ℓ_1 yields a faster convergence rate than AMP from both empirical and theoretical points of view.

Algorithm 6.1: Tied LAMP-ℓ_1 parameter learning [4]

1: Input $\boldsymbol{B} = \boldsymbol{I}$, $\alpha_0 = 1$, $\beta_0 = 1$
2: Learn $\boldsymbol{\Omega}_0^{\text{tied}} = \{\boldsymbol{B}, \alpha_0\}$
3: for $t = 1$ to $T - 1$ do
4: Initialize $\alpha_t = \alpha_{t-1}$, $\beta_t = \beta_{t-1}$
5: Learn $\{\alpha_t, \beta_t\}$ with fixed $\boldsymbol{\Omega}_{t-1}^{\text{tied}}$
6: Re-learn $\boldsymbol{\Omega}_t^{\text{tied}} = \left\{\boldsymbol{B}, \{\alpha_i, \beta_i\}_{i=1}^t, \alpha_0\right\}$
7: end for
8: Return $\boldsymbol{\Omega}_{T-1}^{\text{tied}}$

Algorithm 6.2: Untied LAMP-ℓ_1 parameter learning [4]

1: Learn $\{\boldsymbol{\Omega}_t^{\text{tied}}\}_{t=1}^{T-1}$ via Algorithm 6.1
2: Initialize $\boldsymbol{B}_0 = \boldsymbol{I}, \alpha_0 = 1, \beta_0 = 1$
3: Learn $\boldsymbol{\Omega}_0^{\text{untied}} = \{\boldsymbol{B}_0, \alpha_0\}$
4: for $t = 1$ to $T-1$ do
5: Initialize $\boldsymbol{B}_t = \boldsymbol{B}_{t-1}, \alpha_t = \alpha_{t-1}, \beta_t = \beta_{t-1}$
6: Learn $\{\boldsymbol{B}_t, \alpha_t, \beta_t\}$ with fixed $\boldsymbol{\Omega}_{t-1}^{\text{untied}}$
7: Set $\boldsymbol{\Omega}_t^{\text{untied}} = \{\boldsymbol{B}_i, \alpha_i, \beta_i\}_{i=0}^{t} \setminus \beta_0$ ("\" denotes the set difference operation)
8: if $\boldsymbol{\Omega}_t^{\text{tied}}$ enjoys better performance than $\boldsymbol{\Omega}_t^{\text{untied}}$ then
9: Replace $\boldsymbol{\Omega}_t^{\text{untied}}$ with $\boldsymbol{\Omega}_t^{\text{tied}}$
10: end if
11: Re-learn $\boldsymbol{\Omega}_t^{\text{untied}}$
12: end for
13: Return $\boldsymbol{\Omega}_{T-1}^{\text{untied}}$

6.3.2 Learned Vector-AMP

The VAMP algorithm illustrated in Algorithm 6.3 has been recently proposed in [19] to address AMP's fragility concerning the matrix $\boldsymbol{A}$. Compared to the original AMP, the VAMP algorithm enjoys lower per-iteration complexity and fewer iterations required to convergence. The procedure of the VAMP algorithm is elaborated in the following.

We begin with the definition of the right-rotationally invariant matrices. For the matrix $\boldsymbol{A} \in \mathbb{R}^{L\times N}$ in compressed sensing (2.3), suppose that

$$\boldsymbol{A} = \boldsymbol{U}\boldsymbol{\Lambda}\boldsymbol{V}^{\mathsf{T}} \tag{6.15}$$

satisfies that $\boldsymbol{s} \in \mathbb{R}_+^r$ where $r \triangleq \operatorname{rank}(\boldsymbol{A})$ contains the positive singular values of $\boldsymbol{A}$, then $\boldsymbol{\Lambda} = \operatorname{diag}(\boldsymbol{s}) \in \mathbb{R}^{r\times r}$, $\boldsymbol{U}^{\mathsf{T}}\boldsymbol{U} = \boldsymbol{I}_r$, and $\boldsymbol{V}^{\mathsf{T}}\boldsymbol{V} = \boldsymbol{I}_r$. The matrix $\boldsymbol{A}$ is *right-rotationally invariant* if $\boldsymbol{V}$ consists of the first r columns of a random matrix uniformly distributed on the group of $n \times n$ orthogonal matrices. With any random orthogonal $\boldsymbol{U}$ and a particular distribution on $\boldsymbol{s}$, i.i.d. Gaussian matrices are right-rotationally invariant. The paper [19] demonstrates that with large enough dimensions m, n, VAMP behaves well when the sensing matrix $\boldsymbol{A}$ in compressed sensing is an i.i.d. Gaussian matrix.

The VAMP algorithm consists of two stages which endow with different estimators: LMMSE stage with the estimator

$$\tilde{\boldsymbol{\eta}}(\tilde{\boldsymbol{r}}_t; \tilde{\sigma}_t, \hat{\theta}) := \boldsymbol{V}\left(\operatorname{diag}(\boldsymbol{s})^2 + \frac{\sigma^2}{\tilde{\sigma}_t^2}\boldsymbol{I}_R\right)^{-1}\left(\operatorname{diag}(\boldsymbol{s})\boldsymbol{U}^{\mathsf{T}}\boldsymbol{y} + \frac{\sigma^2}{\tilde{\sigma}_t^2}\boldsymbol{V}^{\mathsf{T}}\tilde{\boldsymbol{r}}_t\right), \tag{6.16}$$

where σ is the standard deviation, and the parameter $\hat{\theta}$ is given by

$$\hat{\theta} := \{\boldsymbol{U}, \boldsymbol{s}, \boldsymbol{V}, \sigma\}, \tag{6.17}$$

Algorithm 6.3: Vector-AMP [19]

Require: LMMSE estimator $\tilde{\eta}(\cdot; \tilde{\sigma}, \hat{\theta})$ (6.16), shrinkage $\eta(\cdot; \sigma, \omega)$ (6.18), max iteration T, parameters $\{\omega_t\}_{t=1}^T$ and $\hat{\theta}$.
1: Set initial $\tilde{r}_1$ and $\tilde{\sigma}_1 > 0$.
2: for $t = 1, 2, \ldots, T$ do
3: // LMMSE stage:
4: $\tilde{x}_t = \tilde{\eta}(\tilde{r}_t; \tilde{\sigma}_t, \hat{\theta})$ // estimation
5: $\tilde{u}_t = \langle \tilde{\eta}'(\tilde{r}_t; \tilde{\sigma}_t, \hat{\theta}) \rangle$ // divergence computation
6: $r_t = (\tilde{x}_t - \tilde{u}_t \tilde{r}_t)/(1 - \tilde{u}_t)$ // Onsager correction
7: $\sigma_t^2 = \tilde{\sigma}_t^2 \tilde{u}_t/(1 - \tilde{u}_t)$ // variance computation
8: // Shrinkage stage:
9: $\hat{x}_t = \eta(r_t; \sigma_t, \omega_t)$ // estimation
10: $u_t = \langle \eta'(r_t, \sigma_t, \omega_t) \rangle$ // divergence computation
11: $\tilde{r}_{t+1} = (\hat{x}_t - u_t r_t)/(1 - u_t)$ // Onsager correction
12: $\tilde{\sigma}_{t+1}^2 = \sigma_t^2 u_t/(1 - u_t)$ // variance computation
13: end for
14: Return $\hat{x}_T$.

and a shrinkage stage with the estimator

$$\eta(r_t; \sigma_t, \alpha) = \eta_{st}(r_t; \alpha\sigma_t), \tag{6.18}$$

where $\eta_{st}(\cdot, \cdot)$ is given by (6.12b). Lines 5 and 10 in Algorithm 6.3 compute the average of the diagonal entries of the Jacobian of $\tilde{\eta}(\cdot; \tilde{\sigma}_t, \hat{\theta})$ and $\eta(\cdot; \sigma_t, \omega_t)$, respectively, which can be referred to [4] for detailed representation.

Based on VAMP illustrated in Algorithm 6.3, we move to the learned VAMP (LVAMP) algorithm proposed in [4]. The t-th layer of the learned VAMP (LVAMP) network consists of four stages: (1) LMMSE optimization, (2) decoupling, (3) shrinkage, and (4) decoupling. The learnable parameters in the t-th layer are the LMMSE stage parameters $\hat{\omega}_t = \{U_t, s_t, V_t, \sigma_t^2\}$ (6.17) and the shrinkage parameters ω_t. Similar to VAMP, the network parameters of LVAMP are concerned in two cases: "tied" and "untied." In the tied case, the network parameters are $\{\hat{\theta}, \{\omega_t\}_{t=1}^T\}$, while in the untied case, it is $\{\hat{\theta}_t, \omega_t\}_{t=1}^T$. Algorithms 6.1 and 6.2 can be exploited to learn the LVAMP parameters for the tied case and the untied case, respectively (with $\hat{\theta}_t$ replacing B_t and with θ_t replacing $\{\alpha_t, \beta_t\}$).

6.3.3 Learned ISTA for Group Row Sparsity

Recall the sparse linear model (2.8) in the multiple-antenna scenario discussed in Sect. 2.2.2, represented by

$$Y = SX + Z, \tag{6.19}$$

where $Y \in \mathbb{C}^{L \times M}$, $X \in \mathbb{C}^{N \times M}$ endowed with group row sparsity.

Algorithm 6.4: ISTA-GS

Require: $\tilde{S}, \tilde{Y}, \lambda$, Iterations.
ISTA-GS($\tilde{S}, \tilde{Y}, \tilde{X}, \lambda$, Iterations):
1: Initialize: $\tilde{X} \leftarrow \mathbf{0}$, $C \leftarrow$ largest eigenvalue of $\tilde{S}^T\tilde{S}$
2: for $i = 1$ to Iterations do
3: $\tilde{X} = \eta_{\lambda/C}(\tilde{X} + \frac{1}{C}\tilde{S}^T(\tilde{Y} - \tilde{S}\tilde{X}))$
4: end for
5: return $\tilde{X}$.

Recall the real-valued counterpart (2.13), and the real-valued counterpart of (6.19) can be represented as

$$\begin{aligned}\tilde{Y} &= \tilde{S}\tilde{X} + \tilde{Z} \\ &= \begin{bmatrix} \Re\{\tilde{S}\} & -\Im\{\tilde{S}\} \\ \Im\{\tilde{S}\} & \Re\{\tilde{S}\} \end{bmatrix} \begin{bmatrix} \Re\{\tilde{X}\} \\ \Im\{\tilde{X}\} \end{bmatrix} + \begin{bmatrix} \Re\{\tilde{Z}\} \\ \Im\{\tilde{Z}\} \end{bmatrix}.\end{aligned} \tag{6.20}$$

The following problem can be established to estimate the group sparse X:

$$\underset{\tilde{X} \in \mathbb{R}^{2N \times M}}{\text{minimize}} \ \|\tilde{Y} - \tilde{S}\tilde{X}\|_F^2 + \lambda \mathscr{R}(\tilde{X}). \tag{6.21}$$

To solve problem (6.21), we start with the ISTA for group row sparse (*ISTA-GS*) illustrated in Algorithm 6.4. Specifically, in the k-th iteration, the update rule is represented as

$$\tilde{X}^{k+1} = \eta_{\lambda/C}\left(\tilde{X}^k + \frac{1}{C}\tilde{S}^T(\tilde{Y} - \tilde{S}\tilde{X}^k)\right), \tag{6.22}$$

where C is the largest eigenvalue of matrix $\tilde{S}^T\tilde{S}$, and $\eta_\theta(\boldsymbol{x}^n)$ denotes the group soft-thresholding function for the n-th row in matrix X (i.e., $\boldsymbol{x}^n$) [2]. Specifically, $\eta_\theta(\boldsymbol{x}^n)$ is defined as

$$\eta_\theta(\boldsymbol{x}^n) = \max\left\{0, \frac{\|\boldsymbol{x}\|_2 - \theta}{\|\boldsymbol{x}\|_2}\right\} \boldsymbol{x}^n. \tag{6.23}$$

Such an iterative algorithm takes a large number of iterations to converge. To address this issue, we present the LISTA-GS method, which parameterizes the iterative method.

Inspired by [5, 9] and by denoting $W_1 = \frac{1}{C}\tilde{S}$, $W_2 = I - \frac{1}{C}\tilde{S}^T\tilde{S}$, and $\theta = \frac{\lambda}{C}$, we rewrite (6.22) as

$$\tilde{X}^{k+1} = \eta_{\theta^k}(W_1\tilde{Y} + W_2\tilde{X}^k). \tag{6.24}$$

The key idea of the proposed LISTA-GS method is to view matrix $\boldsymbol{W_1}$, matrix $\boldsymbol{W_2}$, and scalar θ^k in (6.24) as trainable parameters. As a result, (6.24) can be modeled as a one-layer RNN. Moreover, the unrolled RNN with K iterations for group row sparse can be expressed as

$$\tilde{\boldsymbol{X}}^{k+1} = \eta_{\theta^k}(\boldsymbol{W}_1^k\tilde{\boldsymbol{Y}} + \boldsymbol{W}_2^k\tilde{\boldsymbol{X}}^k), k = 0, 1, \ldots, K-1, \tag{6.25}$$

where all parameters $\boldsymbol{\Theta} = \{\boldsymbol{W}_1^k, \boldsymbol{W}_2^k, \theta^k\}_{k=0}^{K-1}$ are trainable. This is a main difference from the problem formulation in (6.22).

6.3.3.1 Simulations Results

In this section, we present the simulation results of the LISTA-GS method for the joint device activity detection and channel estimation, and compare the results with that of the ISTA-GS method.

In simulations, we generate the signature matrix according to the complex Gaussian distribution, i.e., $\boldsymbol{S} \sim \mathcal{CN}(\boldsymbol{0}, \boldsymbol{I})$. The channels are assumed to suffer from independent Rayleigh fading, i.e., $\boldsymbol{H} \sim \mathcal{CN}(\boldsymbol{0}, \boldsymbol{I})$. In addition, we set the length of the signature sequence (L), the total number of devices (N), and the number of antennas at the BS (M) as 100, 200, and 2, respectively. Each entry of the activity sequence $\{a_1, \ldots, a_N\}$ follows the Bernoulli distribution with mean $p = 0.1$, i.e., $\mathbb{P}(a_n = 1) = 0.1$ and $\mathbb{P}(a_n = 0) = 0.9$, $\forall n \in \mathcal{N}$. After normalizing $\boldsymbol{S}$, we transform all these complex-valued matrices into real-value matrices according to (6.20). Hence, we obtain the training data set $\{\tilde{\boldsymbol{X}}_i^*, \tilde{\boldsymbol{Y}}_i\}_{i=1}^N$. In the training stage, the batch size is set to be 64 and the validation set contains 1000 samples. The learning rate is set to be 10^{-3}. In the testing stage, 1000 samples are generated to test the trained LISTA-GS model. As for ISTA-GS, we set $\lambda = 0.2, 0.1$, and 0.05. We choose $K = 16$ layers for the LISTA-GS method in all the simulations. Furthermore, we initialize the parameters as $W_1^0 = \frac{1}{C}\tilde{\boldsymbol{S}}$, $W_2^0 = \boldsymbol{I} - \frac{1}{C}\tilde{\boldsymbol{S}}^T\tilde{\boldsymbol{S}}$, and $\theta = \frac{0.1}{C}$.

We adopt the normalized mean square error (NMSE) to evaluate the performance of LISTA-GS and ISTA-GS in recovering the real-valued $\tilde{\boldsymbol{X}}$, defined as

$$\text{NMSE}(\tilde{\boldsymbol{X}}, \tilde{\boldsymbol{X}}^*) = 10\log_{10}\left(\frac{\mathbb{E}\|\tilde{\boldsymbol{X}} - \tilde{\boldsymbol{X}}^*\|_F^2}{\mathbb{E}\|\tilde{\boldsymbol{X}}^*\|_F^2}\right), \tag{6.26}$$

where $\tilde{\boldsymbol{X}}^*$ represents the ground truth and $\tilde{\boldsymbol{X}}$ is the estimate obtained by the ISTA-GS and LISTA-GS methods.

As suggested in [5], we train the LISTA-GS model by adopting the layer-wise training strategy, which has been widely used in the previous ISTA models. To stabilize the training process, we add two decayed learning rates, i.e., $\beta_1 = 0.2\beta_0$ and $\beta_2 = 0.02\beta_0$, where β_0 is the initial learning rate. Note that $\boldsymbol{\Theta}^i =$

$\{\boldsymbol{W}_1^k, \boldsymbol{W}_2^k, \theta^k\}_{k=0}^i$ are all the weights from layer 0 to layer i and $m(\cdot)$ is the learning multiplier. We train the RNN layer by layer and the training process of each layer is described as follows:

- Suppose that $\boldsymbol{\Theta}^{i-1}$ is pre-trained for layer i. Initialize the learning multipliers $m(W_1^i), m(W_2^i), m(\theta^i) = 1$.
- Train $\{\boldsymbol{W}_1^i, \boldsymbol{W}_2^i, \theta^i\}$ with β_0.
- Multiply the learning multiplier to their weights and train $\boldsymbol{\Theta}^i = \boldsymbol{\Theta}^{i-1} \cup \{\boldsymbol{W}_1^i, \boldsymbol{W}_2^i, \theta^i\}$ with β_1 and β_2.
- Multiply a decay rate to each learning multiplier.
- Move to train the next layer.

Figure 6.2a shows the NMSE of the proposed LISTA-GS and the baseline ISTA-GS methods over iterations in a noiseless scenario. For the baseline ISTA-GS method, there exists an inherent tradeoff between the convergence rate and the NMSE. In particular, a smaller value of λ results in a more accurate solution but leads to a lower convergence rate, and vice versa. Besides, we observe that LISTA-GS method achieves a much faster convergence rate as well as a much lower NMSE than ISTA-GS for different values of λ. This is because LISTA-GS treats λ as a weight in the training process, yielding a good solution that balances the tradeoff.

Figure 6.2b illustrates the NMSE of the proposed LISTA-GS method over iterations in a noisy scenario with different values of signal-to-noise-ratio (SNR). It can be observed that LISTA-GS can also reach convergence in a few iterations (e.g., less than 16) in the noisy case. As the value of SNR increases, the received power of the pilot sequence increases, which in turn decreases the achievable NMSE.

Figure 6.2c plots the impact of SNR on the NMSE of LISTA-GS and ISTA-GS. The proposed LISTA-GS method achieves a much lower NMSE than ISTA-GS for different values of SNR. In addition, the NMSEs of both methods decrease as the value of SNR increases.

6.4 Summary

In this chapter, we introduced some cutting-edge learning augmented techniques for both structured signal modeling (e.g., structured signal processing under a generative prior [1, 3]) and algorithm design (e.g., learning augmented algorithms [4]). We also introduced Learned ISTA for group row sparsity to solve the sparse linear model in the multiple-antenna scenario. These techniques are summarized in Table 6.1. We hope that this basic idea on learning augmented techniques will provide an intriguing direction for future investigations.

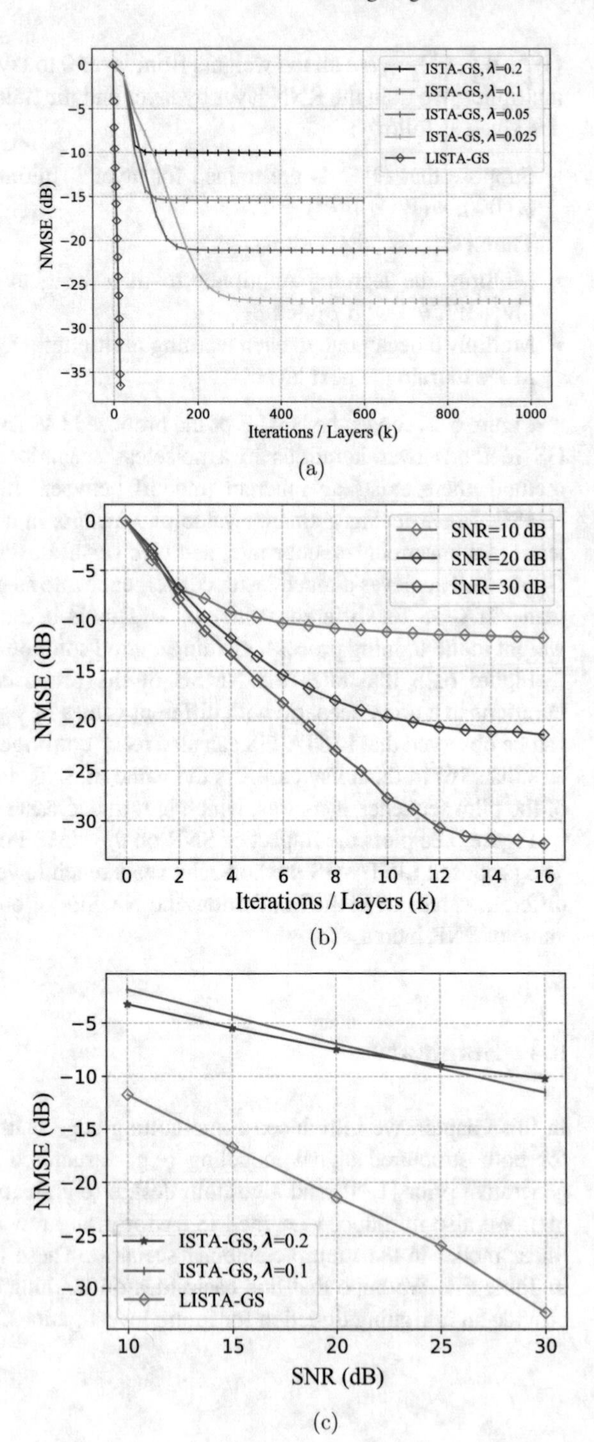

Fig. 6.2 Performance comparison between the proposed LISTA-GS and baseline ISTA-GS in terms of NMSE

Table 6.1 Summary of learning augmented techniques, methods, application, and corresponding references

Techniques	Methods	Application/Reference
Structured signal processing under a generative prior	Learning a generator $G(z): z \in \mathbb{R}^k \to G(z) \in \mathbb{R}^n$	Compressed sensing [3], sparse linear model (2.3), Blind deconvolution [1]
Joint design of measurement matrix and sparse support recovery	Sample scheduling	Compressed sensing [10]
	Sensing matrix optimization	Compressed sensing [6, 7, 14, 15, 24]
	Learning augmented methods: learning an auto-encoder or generative adversarial networks	Linear compression for real signals [18, 22], nonlinear compression for real signals was considered in [17, 23], linear compression for complex signals [13]
Deep-learning-based AMP	Learned AMP	Compressed sensing [4], sparse linear model (2.3), (2.8)
	Learned Vector-AMP	
	Learned ISTA for group sparsity	

References

1. Asim, M., Shamshad, F., Ahmed, A.: Blind image deconvolution using deep generative priors (2018). Preprint. arXiv: 1802.04073
2. Bonnefoy, A., Emiya, V., Ralaivola, L., Gribonval, R.: Dynamic screening: accelerating first-order algorithms for the lasso and group-lasso. IEEE Trans. Signal Process. **63**(19), 5121–5132 (2015)
3. Bora, A., Jalal, A., Price, E., Dimakis, A.G.: Compressed sensing using generative models. In: Proceedings of the International Conference on Machine Learning (ICML), pp. 537–546 (2017). JMLR. org
4. Borgerding, M., Schniter, P., Rangan, S.: AMP-inspired deep networks for sparse linear inverse problems. IEEE Trans. Signal Process. **65**(16), 4293–4308 (2017)
5. Chen, X., Liu, J., Wang, Z., Yin, W.: Theoretical linear convergence of unfolded ISTA and its practical weights and thresholds. In: Advances in Neural Information Processing Systems (NeurIPS), pp. 9061–9071 (2018)
6. Duarte-Carvajalino, J.M., Sapiro, G.: Learning to sense sparse signals: simultaneous sensing matrix and sparsifying dictionary optimization. IEEE Trans. Image Process. **18**(7), 1395–1408 (2009)
7. Elad, M.: Optimized projections for compressed sensing. IEEE Trans. Signal Process. **55**(12), 5695–5702 (2007)
8. Goodfellow, I., Pouget-Abadie, J., Mirza, M., Xu, B., Warde-Farley, D., Ozair, S., Courville, A., Bengio, Y.: Generative adversarial nets. In: Proceedings of the Neural Information Processing Systems (NeurIPS), pp. 2672–2680 (2014)

9. Gregor, K., LeCun, Y.: Learning fast approximations of sparse coding. In: Proceedings of the International Conference on Machine Learning (ICML), pp. 399–406. Omnipress, Madison (2010)
10. Hao, J., Zhang, B., Jiao, Z., Mao, S.: Adaptive compressive sensing based sample scheduling mechanism for wireless sensor networks. Pervasive Mob. Comput. **22**, 113–125 (2015)
11. Kingma, D.P., Ba, J.: ADAM: a method for stochastic optimization. In: Proceedings of the International Conference on Learning Representations (ICLR) (2015)
12. Kingma, D.P., Welling, M.: Auto-encoding variational Bayes (2013). Preprint. arXiv: 1312.6114
13. Li, S., Zhang, W., Cui, Y., Cheng, H.V., Yu, W.: Joint design of measurement matrix and sparse support recovery method via deep auto-encoder. Preprint. arXiv: 1910.04330 (2019)
14. Mangia, M., Rovatti, R., Setti, G.: Rakeness in the design of analog-to-information conversion of sparse and localized signals. IEEE Trans. Circuits Syst. I, Reg. Papers **59**(5), 1001–1014 (2012)
15. Mangia, M., Pareschi, F., Rovatti, R., Setti, G.: Adaptive matrix design for boosting compressed sensing. IEEE Trans. Circuits Syst. I, Reg. Papers **65**(3), 1016–1027 (2017)
16. Mousavi, A., Baraniuk, R.G.: Learning to invert: signal recovery via deep convolutional networks. In: Proceedings of the IEEE International Conference on Acoustics Speech Signal Processing (ICASSP), pp. 2272–2276. IEEE, Piscataway (2017)
17. Mousavi, A., Dasarathy, G., Baraniuk, R.G.: DeepCodec: adaptive sensing and recovery via deep convolutional neural networks. In: Annual Allerton Conference on Communication, Control, and Computing (Allerton), pp. 744–744. IEEE, Piscataway (2017)
18. Mousavi, A., Dasarathy, G., Baraniuk, R.G.: A data-driven and distributed approach to sparse signal representation and recovery. In: Proceedings of the International Conference on Learning Representations (ICLR) (2019)
19. Rangan, S., Schniter, P., Fletcher, A.K.: Vector approximate message passing. IEEE Trans. Inf. Theory **65**, 6664–6684 (2019)
20. Taha, A., Alrabeiah, M., Alkhateeb, A.: Enabling large intelligent surfaces with compressive sensing and deep learning (2019). Preprint. arXiv: 1904.10136
21. Tibshirani, R.: Regression shrinkage and selection via the lasso. J. R. Stat. Soc. **58**(1), 267–288 (1996)
22. Wu, S., Dimakis, A., Sanghavi, S., Yu, F., Holtmann-Rice, D., Storcheus, D., Rostamizadeh, A., Kumar, S.: Learning a compressed sensing measurement matrix via gradient unrolling. In: Proceedings of the International Conference on Machine Learning (ICML), pp. 6828–6839 (2019)
23. Wu, Y., Rosca, M., Lillicrap, T.: Deep compressed sensing. In: Proceedings of the International Conference on Machine Learning (ICML), pp. 6850–6860 (2019)
24. Xu, J., Pi, Y., Cao, Z.: Optimized projection matrix for compressive sensing. EURASIP J. Adv. Signal Process. **2010**(1), 560349 (2010)
25. Yao, S., Zhao, Y., Zhang, A., Su, L., Abdelzaher, T.: DeepIoT: compressing deep neural network structures for sensing systems with a compressor-critic framework. In: Proceedings of the 15th ACM Conference on Embedded Network Sensor Systems, p. 4. ACM, New York (2017)

Chapter 7
Conclusions and Discussions

Abstract This chapter concludes the monograph. A summary is first provided for the main results of each chapter, and two reference tables are provided that contain the main analytical results and algorithms. Furthermore, we provide discussions on the future research directions of low-overhead communications and the corresponding structured signal processing approaches.

7.1 Summary

This monograph investigated different structured signal processing approaches to support low-overhead communications in IoT networks. Chapter 1 provided some background for low-overhead communications in IoT networks, introducing three main techniques that exclude particular parts of the metadata: grant-free random access, pilot-free communications, and identification-free communications. Furthermore, four general structured signal processing models, i.e., a sparse linear model, blind demixing, and sparse blind demixing, a shuffled linear regression, were introduced. Chapters 2–5 formed the core of this monograph, where four general structured signal processing models with corresponding applications in low-overhead communications were presented. For each chapter, the corresponding analysis results and algorithms were provided, which are summarized in Tables 7.1 and 7.2. More details on the proof of analysis results can be referred to Chap. 8. In Chap. 6, some cutting-edge learning augmented based techniques for structured signal processing were introduced, which represent an interesting direction for future research.

Y. Shi et al., *Low-overhead Communications in IoT Networks*,
https://doi.org/10.1007/978-981-15-3870-4_7

Table 7.1 Analytical results in Chaps. 2–6

Result	Description
Theorem 2.1	Approximate kinematic formula captures a phase transition on whether the two randomly rotated cones share a ray
Proposition 2.1	Statistical dimension bound for the smoothed regularizer $\tilde{\mathscr{R}}_G$ (2.22)
Theorem 3.1	The least value of sample size required for exact recovery of problem (3.21)
Theorem 3.2	The convergence analysis of the regularized Wirtinger flow algorithm with spectral initialization for solving the blind demixing problem (3.22)
Theorem 3.3	The convergence analysis of the regularization-free Wirtinger flow algorithm with spectral initialization for solving the blind demixing problem (3.28)
Theorem 3.4	The convergence analysis of the Riemannian gradient with spectral initialization for solving the blind demixing problem (3.41)
Theorem 5.4	The recovery guarantees of Algorithm 5.3 when the shuffled data come from the Gaussian measurement model (5.8)
Theorem 5.5	Demonstrate that the system of polynomial equation (5.27) for exact data has a finite number of solutions, providing theoretical guarantees for developing algebraical method to solving the shuffled linear regression problem
Theorem 5.6	Demonstrate that the system of polynomial equation (5.28) for corrupted data has a finite number of solutions, providing theoretical guarantees for developing algebraical method to solving the shuffled linear regression

Table 7.2 Algorithms in Chaps. 2–6

Algorithm	Description
Algorithm 2.1	Lan, Lu, and Monteiro's
Algorithm 3.1	Initialization via spectral method and projection
Algorithm 3.2	Riemannian optimization on product manifolds
Algorithm 3.3	Riemannian gradient descent with spectral initialization
Algorithm 4.1	DC algorithm for the sparse blind demixing problem (4.18)
Algorithm 5.1	Exact algorithm for calculating the maximum likelihood estimate of the permutation, i.e., (5.12)
Algorithm 5.2	"Row Sampling" algorithm that is exploited as the initialization of Algorithm 5.3
Algorithm 5.3	Approximation algorithm for computing the maximum likelihood estimator (5.9) for shuffled linear regression
Algorithm 5.4	Algebraically-initialized expectation-maximization for shuffled linear regression
Algorithm 6.1	Tied LAMP-ℓ_1 parameter learning for solving lasso (6.1)
Algorithm 6.2	Untied LAMP-ℓ_1 parameter learning for solving lasso (6.1)
Algorithm 6.3	Vector AMP for solving lasso (6.1)
Algorithm 6.4	Learned ISTA for the sparse linear model endowed with group row sparsity, i.e., (6.19)

7.2 Discussions

From Fig. 1.1 which illustrates an exemplary packet structure, we observe other opportunities to further reduce the overhead of the packet. For instance, for device activity detection or blind demixing, a smaller pilot length is preferred. According to recent studies [2, 3], generative models can yield a much more precise representation of the sparse signals. That is, much fewer measurements are required for the recovery of structural signal processing under a generative prior, compared with traditional analytical models. These methods are known as learning augmented methods, some of which were introduced in Chap. 6. They provide a promising direction for future study. Moreover, from the structured signal processing point of view, more sophisticated models are expected to be proposed via exploiting the sporadic activity pattern in massive connectivity networks, by exploiting spatial and temporal correlation of device activities. Furthermore, both convex methods and nonconvex methods with theoretical guarantees have been evoking researcher's interests. For instance, the rigorous statistical analysis for sparse blind demixing is called for further investigation. This is more challenging than the state-of-the-art model [1, 5] which assumes that $\{\boldsymbol{x}_i\}$ in

$$y_j = \sum_{i=1}^{s} \boldsymbol{b}_j^{\mathsf{H}} \boldsymbol{h}_i \boldsymbol{x}_i^{\mathsf{H}} \boldsymbol{a}_{ij}, \ 1 \le j \le L,$$

are sparse. From the algorithmic perspective, more efficient and robust algorithms are expected to be developed. For example, the learning argument methods have been exploited to solve the sparse linear model, e.g., Learned Vector-AMP, Learned AMP [4], and Learned ISTA for group sparsity, as introduced in Sect. 6.3.3.

Overall, given the promising results in applying structured signal processing for low-overhead communications reported in this monograph, we expect these methods will find abundant applications in practical IoT networks. We also hope the methods introduced in the monograph will lead to more effective algorithms, and inspire innovative approaches to exploit various structures in IoT systems, thus enable more application scenarios.

References

1. Ahmed, A., Demanet, L.: Leveraging diversity and sparsity in blind deconvolution. IEEE Trans. Inf. Theory **64**(6), 3975–4000 (2018)
2. Asim, M., Shamshad, F., Ahmed, A.: Blind image deconvolution using deep generative priors. arXiv preprint:1802.04073 (2018)
3. Bora, A., Jalal, A., Price, E., Dimakis, A.G.: Compressed sensing using generative models. In: Proceedings of the 34th International Conference on Machine Learning (ICML), vol. 70, pp. 537–546 (2017). JMLR.org

4. Borgerding, M., Schniter, P., Rangan, S.: AMP-inspired deep networks for sparse linear inverse problems. IEEE Trans. Signal Process. **65**(16), 4293–4308 (2017)
5. Flinth, A.: Sparse blind deconvolution and demixing through $\ell_{1,2}$-minimization. Adv. Comput. Math. **44**(1), 1–21 (2018)

Chapter 8
Appendix

8.1 Conic Integral Geometry

In this section, several basic concepts of conic integral geometry theory are introduced. We begin with the kinematic formula for cones which is the probability that a randomly rotated convex cone shares a ray with a fixed convex cone. This formula plays a vital role in characterizing the success or failure probability of an estimation problem. The following introduction is based on [3].

8.1.1 *The Kinematic Formula for Convex Cones*

In the area of conic integral geometry, it is critical to identify the probability that a randomly rotated convex cone shares a ray with a fixed convex cone. Considering convex cones C and S in $\mathbb{R}^d$, and a random orthogonal basis $\boldsymbol{A} \in \mathbb{R}^{d\times d}$, we aim to find an effective expression for the probability

$$P\{C \cap \boldsymbol{A}S \neq \{\boldsymbol{0}\}\}. \tag{8.1}$$

Studying this probability enables to understand the phase transition phenomena in convex optimization problems with random data.

We start with the simple case of two dimensions where the solution to the problem can be quickly computed. Consider two convex cones C and S in $\mathbb{R}^2$, and assume that neither cone is a linear subspace. Then

$$P\{C \cap \boldsymbol{A}S \neq \{\boldsymbol{0}\}\} = \min\{\upsilon_2(C) + \upsilon_2(S),\ 1\}, \tag{8.2}$$

where $\upsilon_2(\cdot)$ returns the portion of the unit circle united by a convex cone in $\mathbb{R}^2$. If one of the cones is a subspace, a similar formula can be derived. In spaces with

Y. Shi et al., *Low-overhead Communications in IoT Networks*,
https://doi.org/10.1007/978-981-15-3870-4_8

higher dimensions, the representation of convex cones becomes more complicated. In three dimensions, it might be troublesome to find a reasonable solution in general. To address this issue, an extraordinary tool called the *conic kinematic formula* [19, Thm. 6.5.6] has been developed. It shows that there exists an *exact* formula to identify the probability that a randomly rotated convex cone shares a ray with a fixed convex cone. Moreover, only $d + 1$ numbers are needed to summarize each cone in d dimensions.

Fact 8.1 (The Kinematic Formula for Cones) *Let C and S be closed convex cones in $\mathbb{R}^n$, one of which is not a subspace. Assuming a random orthogonal basis $\boldsymbol{A} \in \mathbb{R}^{n\times n}$, then*

$$P\{C \cap \boldsymbol{A}S \neq \{\boldsymbol{0}\}\} = \sum_{i=0}^{n} \left(1 + (-1)^{i+1}\right) \sum_{j=i}^{n} \upsilon_i(C) \cdot \upsilon_{n+i-j}(S). \tag{8.3}$$

For each $k = 0, 1, 2, \ldots, n$, the operation υ_k maps a closed convex cone to a nonnegative number, called the k-th intrinsic volume *of the cone.*

Even though the conic kinematic formula is beneficial for studying random instances of convex optimization problems [2, 15], this approach suffers a strenuous computation of expressions for the intrinsic volumes of a cone, except in the simplest cases. To address this issue, the paper [3] provided a novel method that makes the kinematic formula effective, which is elaborated in the following.

8.1.2 Intrinsic Volumes and the Statistical Dimension

The conic intrinsic volumes, illustrated in Fact 8.1, are the elemental geometric invariants of a closed convex cone. That is, the conic intrinsic volumes do not depend on the orientation of the cone within the space in which the cone is embedded, nor on the dimension of that space. This quantity is similar to some quantity defined for compact convex sets in Euclidean geometry, such as the usual volume, the surface area, the mean width, and the Euler characteristic [18].

The intrinsic volume of a closed convex cone C in $\mathbb{R}^n$ consists of a sequel of probability distributions on $\{0, 1, 2, \ldots, n\}$, represented as

$$\sum_{i=0}^{n} \upsilon_i(C) = 1 \quad \text{and} \quad \upsilon_i(C) \geq 0 \quad \text{for } i = 0, 1, 2, \ldots, n. \tag{8.4}$$

The work [3] established an extraordinary fact about conic geometry: for each closed convex cone, the distribution of conic intrinsic volumes sharply concentrates around its mean value. The precise statement on the concentration of intrinsic is illustrated in Theorem 8.1. To begin with, we introduce several definitions that contribute to Theorem 8.1.

Definition 8.1 (Tail Functionals) Let C be a closed convex cone in $\mathbb{R}^n$. For every $k = 0, 1, 2, \dots, n$, the k-th *tail functional* is defined as

$$t_k(C) := \upsilon_k(C) + \upsilon_{k+1}(C) + \cdots = \sum_{j=k}^{n} \upsilon_j(C). \tag{8.5}$$

The k-th *half-tail functional* is given by

$$h_k(C) := \upsilon_k(C) + \upsilon_{k+2}(C) + \cdots = \sum_{\substack{j=k \\ j-k \text{ even}}}^{n} \upsilon_j(C). \tag{8.6}$$

Definition 8.2 (Statistical Dimension) Let C be a closed convex cone in $\mathbb{R}^n$. The *statistical dimension* $\delta(C)$ of the cone is given by

$$\delta(C) := \sum_{k=0}^{n} k\, \upsilon_k(C). \tag{8.7}$$

As Definition 8.1 shows, the statistical dimension indicates the dimensionality of a convex cone. In particular, the statistical dimension is a *canonical extension* of the dimension of a linear subspace to the class of convex cones.

Based on the aforementioned definition, we arrive at the theorem that demonstrates the concentration of intrinsic volumes.

Theorem 8.1 (Concentration of Intrinsic Volumes [3]) *Assuming that C is a closed convex cone, the transition width is given by*

$$\rho(C) := \sqrt{\delta(C^\circ) \wedge \delta(C)}.$$

Define a function

$$p_C(\gamma) := 4 \exp\left(\frac{-\gamma^2/8}{\rho^2(C) + \gamma} \right) \quad \textit{for } \gamma \geq 0. \tag{8.8}$$

Then

$$k_- \leq \delta(C) - \gamma + 1 \quad \Longrightarrow \quad t_{k_-}(C) \geq 1 - p_C(\gamma); \tag{8.9}$$

$$k_+ \geq \delta(C) + \gamma \quad \Longrightarrow \quad t_{k_+}(C) \leq p_C(\gamma), \tag{8.10}$$

where t_k (8.5) is the tail functional, and $\wedge$ is the operator that returns the minimum of two numbers.

8.1.3 The Approximate Kinematic Formula

Based on the concentration of intrinsic volumes provided in Theorem 8.1 and the conic kinematic formula illustrated in Fact 8.1, we can arrive at the following approximate kinematic formula.

Theorem 8.2 (Approximate Kinematic Formula [3]) *Define a fix parameter* $\alpha \in (0, 1)$. *Let* C *and* S *be convex cones in* $\mathbb{R}^n$, *and assume a random orthogonal basis* $\boldsymbol{A} \in \mathbb{R}^{n\times n}$. *Then*

$$\delta(C) + \delta(S) \leq n - \sqrt{n8\log(4/\alpha)} \quad \Longrightarrow \quad P\{C \cap \boldsymbol{A}S \neq \{\boldsymbol{0}\}\} \leq \alpha;$$

$$\delta(C) + \delta(S) \geq n + \sqrt{n8\log(4/\alpha)} \quad \Longrightarrow \quad P\{C \cap \boldsymbol{A}S \neq \{\boldsymbol{0}\}\} \geq 1 - \alpha.$$

Theorem 8.2 demonstrates that two rotated cones are prone to share a ray in the case that the total statistical dimension of the two cones exceeds the ambient dimension. For problems in conic integral geometry, the cone is analogous to a subspace with approximate dimension $\delta(C)$. In the paper [3], a large class of random convex optimization problems have been proved to exhibit a phase transition, and the statistical dimension corresponding to each convex optimization problem characterizes the location of the phase transition.

8.1.4 Computing the Statistical Dimension

The statistical dimension plays a vital role in conic integral geometry, which can be used to identify that phase transitions occur in random convex optimization problems. To efficiently compute the statistical dimension, the method proposed in [3] is presented in the follows. We begin with several basic definitions. For a closed convex cone C, the projection $\text{Proj}_C(\boldsymbol{x})$ that maps a point $\boldsymbol{x}$ onto a point on the cone C which is nearest to $\boldsymbol{x}$:

$$\boldsymbol{\Pi}_C(\boldsymbol{x}) := \text{argmin}\left\{\|\boldsymbol{x} - \boldsymbol{y}\| : \boldsymbol{y} \in C\right\}. \tag{8.11}$$

For a general cone $C \subset \mathbb{R}^n$, the *polar cone* C° is defined as the set of outward normals of C:

$$C^\circ := \left\{\boldsymbol{y} \in \mathbb{R}^n : \langle \boldsymbol{y}, \boldsymbol{x}\rangle \leq 0 \quad \text{for all } \boldsymbol{x} \in C\right\}. \tag{8.12}$$

Proposition 8.1 (Statistical Dimension (Recall Definition 2.2)) *The statistical dimension* $\delta(C)$ *of a closed convex cone* C *in* $\mathbb{R}^n$ *satisfies*

$$\delta(C) = \mathbb{E}\left[\,\|\boldsymbol{\Pi}_C(\boldsymbol{g})\|_2\,\right], \tag{8.13}$$

where $\boldsymbol{g} \in \mathbb{R}^d$ *is a standard Gaussian vector, and* $\boldsymbol{\Pi}_C$ *is defined in* (8.11).

The metric characterization of the statistical dimension illustrated in Proposition 8.1 enables to connect the approach based on integral geometry and to the approach based on Gaussian process theory. The results can be obtained by a classic argument called the spherical Steiner formula [19, Thm. 6.5.1]. Furthermore, the formula (8.13) is related to another definition of parameter for convex cones called the *Gaussian width*, i.e., for a convex cone $C \subset \mathbb{R}^n$, the width is defined as

$$w(C) := \mathbb{E}\big[\sup_{\boldsymbol{y}\in C\cap \mathbb{S}^{n-1}} \langle \boldsymbol{y},\ \boldsymbol{g}\rangle\big],$$

where $\boldsymbol{g} \in \mathbb{R}^d$ is a standard Gaussian vector. This relation enables us to compute the statistical dimension by exploiting methods [4, 17] developed for the Gaussian width.

8.2 Proof of Proposition 2.1

Without loss of generality, define that

$$\boldsymbol{\Theta}_0 = \left[\left(\boldsymbol{\theta}_0^1\right)^T, \ldots, \left(\boldsymbol{\theta}_0^S\right)^\top, \boldsymbol{0}_{M\times(N-S)}\right]^T \in \mathbb{C}^{N\times M},$$

where $\boldsymbol{\theta}_0^i$ are nonzero. Hence, (2.21) is reformulated as

$$\delta\left(\mathscr{D}\left(\tilde{\mathscr{R}}_G;\tilde{\boldsymbol{\Theta}}_0\right)\right) \le \inf_{\eta\ge 0} \mathbb{E}\left[\mathrm{dist}^2\left(\boldsymbol{G}, \eta\cdot\partial\tilde{\mathscr{R}}_G\left(\tilde{\boldsymbol{\Theta}}_0\right)\right)\right], \tag{8.14}$$

where $\boldsymbol{G} \in \mathbb{R}^{2N\times M}$ is a standard Gaussian matrix. Since $\partial\tilde{\mathscr{R}}_G(\tilde{\boldsymbol{\Theta}}_0) = \partial\mathscr{R}_G(\tilde{\boldsymbol{\theta}}_0) + \frac{\mu}{2}\partial\|\tilde{\boldsymbol{\Theta}}_0\|_F^2$, we have

$$\begin{aligned}
&\boldsymbol{U} \in \partial\tilde{\mathscr{R}}_G\left(\tilde{\boldsymbol{\Theta}}_0\right)\\
&\Longleftrightarrow \begin{cases} \boldsymbol{U}_{\mathscr{V}_j} = \left(\tilde{\boldsymbol{\Theta}}_0\right)_{\mathscr{V}_j} / \left\|\left(\tilde{\boldsymbol{\Theta}}_0\right)_{\mathscr{V}_j}\right\|_F + \mu\left(\tilde{\boldsymbol{\Theta}}_0\right)_{\mathscr{V}_j} & \text{if } j = 1,\ldots,S,\\ \|\boldsymbol{U}_{\mathscr{V}_j}\|_F \le 1 & \text{if } j = S+1,\ldots,N, \end{cases}
\end{aligned} \tag{8.15}$$

where $\tilde{\boldsymbol{\Theta}}_{\mathscr{V}_j} = \boldsymbol{0}$ for $j \ne i$ for some $\tilde{\boldsymbol{\Theta}} \in \mathbb{R}^{2N\times M}$, defined in (2.23). Hence,

$$\begin{aligned}
\mathrm{dist}^2(\boldsymbol{G}, \eta\cdot\partial\tilde{\mathscr{R}}_G(\tilde{\boldsymbol{\Theta}}_0)) &= \sum_{i=1}^{S} \|\boldsymbol{G}_{\mathscr{V}_i} - \eta((\tilde{\boldsymbol{\Theta}}_0)_{\mathscr{V}_i}/\|(\tilde{\boldsymbol{\Theta}}_0)_{\mathscr{V}_i}\|_F + \mu(\tilde{\boldsymbol{\Theta}}_0)_{\mathscr{V}_j})\|_F^2\\
&\quad + \sum_{i=S+1}^{N} \max\{\|\boldsymbol{G}_{\mathscr{V}_i}\|_2 - \eta, 0\}^2.
\end{aligned} \tag{8.16}$$

Taking the expectation over the Gaussian matrix $\boldsymbol{G}$, it arrives

$$\mathbb{E}\left[\operatorname{dist}^2\left(\boldsymbol{G}, \eta \cdot \partial\mathscr{R}_G\left(\tilde{\boldsymbol{\Theta}}_0\right)\right)\right] = S\left(2M + \eta^2\left(1 + 2\mu\bar{a} + \mu^2\bar{b}\right)\right) + (N-S)\frac{2^{1-M}}{\Gamma(M)}\int_\eta^\infty (u-\eta)^2 u^{2M-1} e^{-\frac{u^2}{2}}\,\mathrm{d}u,$$

where $\bar{a} = \frac{1}{S}\sum_{i=1}^S \|(\tilde{\boldsymbol{\Theta}}_0)_{\mathcal{V}_i}\|_F$ and $\bar{b} = \frac{1}{S}\sum_{i=1}^S \|(\tilde{\boldsymbol{\Theta}}_0)_{\mathcal{V}_i}\|_F^2$. Letting $\rho = S/N$ and taking the infimum over $\eta \geq 0$ completes the proof of (2.24).

8.3 Proof of Theorem 3.3

Theorem 3.3 can be justified via trajectory analysis for blind demixing via the Wirtinger flow algorithm. This is achieved by proving that iterates of Wirtinger flow sustain in the region of incoherence and contraction by exploiting the local geometry of blind demixing. The steps of proving Theorem 3.3 are summarized as follows.

- **Identifying local geometry in the region of incoherence and contraction (RIC).** First identify a region $\mathscr{R}$, i.e., RIC, where the objective function enjoys restricted strong convexity and smoothness near the ground truth $\boldsymbol{z}^\natural$. Furthermore, any point $\boldsymbol{z} \in \mathscr{R}$ obeys the ℓ_2 error contraction and the incoherence conditions. Please refer to Lemma 8.1 for details. Hence, the convergence rate of the algorithm can be established according to Lemma 8.2, if and only if all the iterates of Wirtinger flow with spectral initialization are in the region $\mathscr{R}$.
- **Establishing the auxiliary sequences via the leave-one-out approach.** To justify that the Wirtinger Flow algorithm enforces the iterates to stay within the RIC, we introduce the leave-one-out sequences. Specifically, the leave-one-out sequences are denoted by $\{\boldsymbol{h}_i^{t,(l)}, \boldsymbol{x}_i^{t,(l)}\}_{t\geq 0}$ for each $1 \leq i \leq s$, $1 \leq l \leq m$ obtained by removing the l-th measurement from the objective function $f(\boldsymbol{h}, \boldsymbol{x})$. Hence, $\{\boldsymbol{h}_i^{t,(l)}\}$ and $\{\boldsymbol{x}_i^{t,(l)}\}$ are independent with $\{\boldsymbol{b}_j\}$ and $\{\boldsymbol{a}_{ij}\}$, respectively.
- **Establishing the incoherence condition via induction.** In this step, we employ the auxiliary sequences to establish the incoherence condition via induction. For brief, with $\widetilde{\boldsymbol{z}}_i^t = [\widetilde{\boldsymbol{z}}_1^{t*}, \ldots, \widetilde{\boldsymbol{z}}_s^{t*}]^*$ where $\widetilde{\boldsymbol{z}}_i^t = [\widetilde{\boldsymbol{h}}_i^{t*}\ \widetilde{\boldsymbol{x}}_i^{t*}]^*$, the set of induction hypotheses of local geometry is listed as follows:

$$\operatorname{dist}\left(\boldsymbol{z}^t, \boldsymbol{z}^\natural\right) \leq C_1 \frac{1}{\log^2 m}, \tag{8.17a}$$

$$\operatorname{dist}\left(\boldsymbol{z}^{t,(l)}, \widetilde{\boldsymbol{z}}^t\right) \leq C_2 \frac{s\mu}{\sqrt{m}}\sqrt{\frac{\mu^2 K \log^9 m}{m}}, \tag{8.17b}$$

$$\max_{1\leq i\leq s, 1\leq j\leq m} \left| \boldsymbol{a}_{ij}^* \left(\widetilde{\boldsymbol{x}}_i^t - \boldsymbol{x}_i^\natural \right) \right| \leq C_3 \frac{1}{\sqrt{s}\log^{3/2} m}, \tag{8.17c}$$

$$\max_{1\leq i\leq s, 1\leq j\leq m} \left| \boldsymbol{b}_l^* \widetilde{\boldsymbol{h}}_i^t \right| \leq C_4 \frac{\mu}{\sqrt{m}} \log^2 m, \tag{8.17d}$$

where C_1, C_3 are some sufficiently small constants, while C_2, C_4 are some sufficiently large constants. That is, as long as the current iterate stays within the RIC, the next iterate remains in the RIC.

- **Concentration between original and auxiliary sequences.** The gap between $\{z^t\}$ and $\{z^{t,(l)}\}$ can be established via employing the restricted strong convexity of the objective function in RIC.
- **Incoherence condition of auxiliary sequences.** Based on the fact that $\{z^t\}$ and $\{z^{t,(l)}\}$ are sufficiently close, we can instead bound the incoherence of $\boldsymbol{h}_i^{t,(l)}$ (resp. $\boldsymbol{x}_i^{t,(l)}$) in terms of $\{\boldsymbol{b}_j\}$ (resp. $\{\boldsymbol{a}_{ij}\}$), which turns out to be much easier due to the statistical independence between $\{\boldsymbol{h}_i^{t,(l)}\}$ (resp. $\{\boldsymbol{x}_i^{t,(l)}\}$) and $\{\boldsymbol{b}_j\}$ (resp.$\{\boldsymbol{a}_{ij}\}$).
- **Establishing iterates in RIC.** By combining the above bounds together, we arrive at $|\boldsymbol{a}_{ij}^*(\boldsymbol{x}_i^t - \boldsymbol{x}_i^\natural)| \leq \|\boldsymbol{a}_{ij}\|_2 \cdot \|\boldsymbol{x}_i^t - \boldsymbol{x}_i^{t,(l)}\|_2 + \|\boldsymbol{a}_{ij}^*(\boldsymbol{x}_i^{t,(l)} - \boldsymbol{x}_i^\natural)\|$ via the triangle inequality. Based on the similar arguments, the other incoherence condition can be established in Lemma 8.3.
- **Establishing initial point in RIC.** Lemmas 8.6–8.8 are integrated to justify that the spectral initialization point is in RIC.

Lemma 8.1 (Restricted Strong Convexity and Smoothness for Blind Demixing Problem $\mathscr{P}$) *Let $\delta > 0$ be a sufficiently small constant. If the number of measurements satisfies $m \gg \mu^2 s^2 \kappa^2 K \log^5 m$, then with probability at least $1 - O(m^{-10})$, the Wirtinger Hessian $\nabla^2 f_{\text{clean}}(z)$ obeys*

$$\boldsymbol{u}^* \left[\boldsymbol{D}\nabla^2 f_{\text{clean}}(z) + \nabla^2 f_{\text{clean}}(z)\boldsymbol{D} \right] \boldsymbol{u} \geq \frac{1}{4\kappa}\|\boldsymbol{u}\|_2^2 \text{ and}$$

$$\left\| \nabla^2 f_{\text{clean}}(z) \right\| \leq 2 + s \tag{8.18}$$

simultaneously for all

$$\boldsymbol{u} = \begin{bmatrix} \boldsymbol{u}_1 \\ \vdots \\ \boldsymbol{u}_s \end{bmatrix} \text{ with } \boldsymbol{u}_i = \begin{bmatrix} \boldsymbol{h}_i - \boldsymbol{h}_i' \\ \boldsymbol{x}_i - \boldsymbol{x}_i' \\ \overline{\boldsymbol{h}_i - \boldsymbol{h}_i'} \\ \overline{\boldsymbol{x}_i - \boldsymbol{x}_i'} \end{bmatrix},$$

$$\text{and } \boldsymbol{D} = \operatorname{diag}\left(\{\boldsymbol{W}_i\}_{i=1}^s\right)$$

$$\text{with } \boldsymbol{W}_i = \operatorname{diag}\left(\left[\beta_{i1}\boldsymbol{I}_K \; \beta_{i2}\boldsymbol{I}_K \; \overline{\beta_{i1}}\boldsymbol{I}_K \; \overline{\beta_{i2}}\boldsymbol{I}_K\right]^*\right).$$

Here z satisfies

$$\max_{1\le i\le s}\max\left\{\left\|\boldsymbol{h}_i-\boldsymbol{h}_i^{\natural}\right\|_2,\left\|\boldsymbol{x}_i-\boldsymbol{x}_i^{\natural}\right\|_2\right\}\le\frac{\delta}{\kappa\sqrt{s}}, \tag{8.19a}$$

$$\max_{1\le i\le s,1\le j\le m}\left|\boldsymbol{a}_{ij}^*\left(\boldsymbol{x}_i-\boldsymbol{x}_i^{\natural}\right)\right|\cdot\left\|\boldsymbol{x}_i^{\natural}\right\|_2^{-1}\le\frac{2C_3}{\sqrt{s}\log^{3/2}m}, \tag{8.19b}$$

$$\max_{1\le i\le s,1\le j\le m}|\boldsymbol{b}_j^*\boldsymbol{h}_i|\cdot\left\|\boldsymbol{h}_i^{\natural}\right\|_2^{-1}\le\frac{2C_4\mu}{\sqrt{m}}\log^2 m, \tag{8.19c}$$

where $(\boldsymbol{h}_i,\boldsymbol{x}_i)$ *is aligned with* $(\boldsymbol{h}_i',\boldsymbol{x}_i')$, *and one has* $\max\{\|\boldsymbol{h}_i-\boldsymbol{h}_i^{\natural}\|_2,\|\boldsymbol{h}_i'-\boldsymbol{h}_i^{\natural}\|_2,\|\boldsymbol{x}_i-\boldsymbol{x}_i^{\natural}\|_2,\|\boldsymbol{x}_i'-\boldsymbol{x}_i^{\natural}\|_2\}\le\delta/(\kappa\sqrt{s})$, *for* $i=1,\ldots,s$ *and* $\boldsymbol{W}_i$*'s satisfy that for* $\beta_{i1},\beta_{i2}\in\mathbb{R}$, *for* $i=1,\ldots,s$ $\max_{1\le i\le s}\max\left\{|\beta_{i1}-\frac{1}{\kappa}|,|\beta_{i2}-\frac{1}{\kappa}|\right\}\le\frac{\delta}{\kappa\sqrt{s}}$. *Therein,* $C_3,C_4\ge 0$ *are numerical constants.*

Based on the local geometry in the region of incoherence and contraction, we further establish contraction of the error measured by the distance function.

Lemma 8.2 *Suppose the number of measurements satisfies* $m\gg\mu^2s^2\kappa^2K\log^5 m$ *and the step size obeys* $\eta>0$ *and* $\eta\asymp s^{-1}$. *Then with probability at least* $1-O(m^{-10})$, *we have*

$$\operatorname{dist}\left(\boldsymbol{z}^{t+1},\boldsymbol{z}^{\natural}\right)\le(1-\eta/(16\kappa))\operatorname{dist}\left(\boldsymbol{z}^t,\boldsymbol{z}^{\natural}\right)+3\kappa\sqrt{s}\max_{1\le k\le s}\|\mathscr{A}_k(\boldsymbol{e})\|,$$

provided that

$$\operatorname{dist}\left(\boldsymbol{z}^t,\boldsymbol{z}^{\natural}\right)\le\xi, \tag{8.20a}$$

$$\max_{1\le i\le s,1\le j\le m}\left|\boldsymbol{a}_{ij}^*\left(\widetilde{\boldsymbol{x}}_i^t-\boldsymbol{x}_i^{\natural}\right)\right|\cdot\|\boldsymbol{x}_i^{\natural}\|_2^{-1}\le\frac{2C_3}{\sqrt{s}\log^{3/2}m}, \tag{8.20b}$$

$$\max_{1\le i\le s,1\le j\le m}\left|\boldsymbol{b}_j^*\widetilde{\boldsymbol{h}}_i^t\right|\cdot\left\|\boldsymbol{h}_i^{\natural}\right\|_2^{-1}\le\frac{2C_4\mu}{\sqrt{m}}\log^2 m, \tag{8.20c}$$

for some constants $C_3,C_4>0$ *and a sufficiently small constant* $\xi>0$. *Here,* $\widetilde{\boldsymbol{h}}_i^t$ *and* $\widetilde{\boldsymbol{x}}_i^t$ *are defined as* $\widetilde{\boldsymbol{h}}_i^t=\frac{1}{\alpha_i^t}\boldsymbol{h}_i^t$ *and* $\widetilde{\boldsymbol{x}}_i^t=\alpha_i^t\boldsymbol{x}_i^t$ *for* $i=1,\ldots,s$.

Proof From the definition of α_k^{t+1}, $k = 1, \ldots, s$, one has

$$
\begin{aligned}
\operatorname{dist}\left(z^{t+1}, z^{\natural}\right)^2 &\leq \sum_{k=1}^{s} \operatorname{dist}\left(z_k^{t+1}, z_k^{\natural}\right)^2 \\
&\overset{\text{(i)}}{\leq} s \left\| \frac{1}{\overline{\alpha_k^{t+1}}} \boldsymbol{h}_k^{t+1} - \boldsymbol{h}_k^{\natural} \right\|_2^2 + s \left\| \alpha_k^{t+1} \boldsymbol{x}_k^{t+1} - \boldsymbol{x}_k^{\natural} \right\|_2^2 \\
&\leq s \left\| \frac{1}{\overline{\alpha_k^{t}}} \boldsymbol{h}_k^{t+1} - \boldsymbol{h}_k^{\natural} \right\|_2^2 + s \left\| \alpha_k^{t} \boldsymbol{x}_k^{t+1} - \boldsymbol{x}_k^{\natural} \right\|_2^2,
\end{aligned}
\tag{8.21}
$$

where k in the step (i) satisfies that $k = \arg\max_{1 \leq i \leq s} \operatorname{dist}(z_i^{t+1}, z_i^{\natural})^2$.

By denoting $\widetilde{\boldsymbol{h}}_k^t = \frac{1}{\overline{\alpha_k^t}} \boldsymbol{h}_k^t$, $\widetilde{\boldsymbol{x}}_k^t = \alpha_k^t \boldsymbol{x}_k^t$, $\widehat{\boldsymbol{h}}_k^{t+1} = \frac{1}{\overline{\alpha_k^t}} \boldsymbol{h}_k^{t+1}$ and $\widehat{\boldsymbol{x}}_k^{t+1} = \alpha_k^t \boldsymbol{x}_k^{t+1}$, we have

$$
\begin{bmatrix}
\widehat{\boldsymbol{h}}_k^{t+1} - \boldsymbol{h}_k^{\natural} \\
\widehat{\boldsymbol{x}}_k^{t+1} - \boldsymbol{x}_k^{\natural} \\
\overline{\widehat{\boldsymbol{h}}_k^{t+1} - \boldsymbol{h}_k^{\natural}} \\
\overline{\widehat{\boldsymbol{x}}_k^{t+1} - \boldsymbol{x}_k^{\natural}}
\end{bmatrix}
=
\begin{bmatrix}
\widetilde{\boldsymbol{h}}_k^{t} - \boldsymbol{h}_k^{\natural} \\
\widetilde{\boldsymbol{x}}_k^{t} - \boldsymbol{x}_k^{\natural} \\
\overline{\widetilde{\boldsymbol{h}}_k^{t} - \boldsymbol{h}_k^{\natural}} \\
\overline{\widetilde{\boldsymbol{x}}_k^{t} - \boldsymbol{x}_k^{\natural}}
\end{bmatrix}
- \eta \boldsymbol{W}_k
\begin{bmatrix}
\nabla_{\boldsymbol{h}_k} f(\widetilde{z}^t) \\
\nabla_{\boldsymbol{x}_k} f(\widetilde{z}^t) \\
\overline{\nabla_{\boldsymbol{h}_k} f(\widetilde{z}^t)} \\
\overline{\nabla_{\boldsymbol{x}_k} f(\widetilde{z}^t)}
\end{bmatrix},
\tag{8.22}
$$

where

$$
\boldsymbol{W}_k = \operatorname{diag}\left(\left[\|\widetilde{\boldsymbol{x}}_k^t\|_2^{-2} \boldsymbol{I}_K \quad \left\|\widetilde{\boldsymbol{h}}_k^t\right\|_2^{-2} \boldsymbol{I}_K \quad \|\widetilde{\boldsymbol{x}}_k^t\|_2^{-2} \boldsymbol{I}_K \quad \left\|\widetilde{\boldsymbol{h}}_k^t\right\|_2^{-2} \boldsymbol{I}_K \right] \right).
\tag{8.23}
$$

The Wirtinger Hessian without noise of $f_{\text{clean}}(z)$ in terms of z_i can be written as

$$
\nabla_{z_i}^2 f_{\text{clean}} := \begin{bmatrix} \boldsymbol{C} & \boldsymbol{E} \\ \boldsymbol{E}^* & \overline{\boldsymbol{C}} \end{bmatrix},
\tag{8.24}
$$

where $\boldsymbol{C} := \frac{\partial}{\partial z_i} \left(\frac{\partial f_{\text{clean}}}{\partial z_i} \right)^*$ and $\boldsymbol{E} := \frac{\partial}{\partial \overline{z_i}} \left(\frac{\partial f_{\text{clean}}}{\partial z_i} \right)^*$. The Wirtinger Hessian of $f_{\text{clean}}(z)$ in terms of z is thus represented as

$$
\nabla^2 f_{\text{clean}}(z) := \operatorname{diag}\left(\left\{ \nabla_{z_i}^2 f_{\text{clean}} \right\}_{i=1}^{s} \right),
\tag{8.25}
$$

where the operation $\operatorname{diag}(\{\boldsymbol{A}_i\}_{i=1}^s)$ generates a block diagonal matrix with the diagonal elements being matrices $\boldsymbol{A}_1, \ldots, \boldsymbol{A}_s$. According to the fundamental theorem of calculus provided in [13], together with the definition of the noiseless objective function f_{clean} and the noiseless Wirtinger Hessian $\nabla^2_{z_k} f_{\text{clean}}$, we get $\nabla^2_{z_k} f_{\text{clean}}$,

$$\begin{bmatrix} \nabla_{\boldsymbol{h}_k} f(\widetilde{z}^t) \\ \nabla_{\boldsymbol{x}_k} f(\widetilde{z}^t) \\ \hline \nabla_{\boldsymbol{h}_k} f(\widetilde{z}^t) \\ \hline \nabla_{\boldsymbol{x}_i} f(\widetilde{z}^t) \end{bmatrix} = \begin{bmatrix} \nabla_{\boldsymbol{h}_k} f_{\text{clean}}(\widetilde{z}^t) \\ \nabla_{\boldsymbol{x}_k} f_{\text{clean}}(\widetilde{z}^t) \\ \hline \nabla_{\boldsymbol{h}_k} f_{\text{clean}}(\widetilde{z}^t) \\ \hline \nabla_{\boldsymbol{x}_k} f_{\text{clean}}(\widetilde{z}^t) \end{bmatrix} + \begin{bmatrix} \mathscr{A}_k(\boldsymbol{e})\boldsymbol{x}_k^t \\ \mathscr{A}_k^*(\boldsymbol{e})\boldsymbol{h}_k^t \\ \hline \mathscr{A}_k(\boldsymbol{e})\boldsymbol{x}_k^t \\ \hline \mathscr{A}_k^*(\boldsymbol{e})\boldsymbol{h}_k^t \end{bmatrix}$$

$$= \boldsymbol{H}_k \begin{bmatrix} \widetilde{\boldsymbol{h}}_k^t - \boldsymbol{h}_k^\natural \\ \widetilde{\boldsymbol{x}}_k^t - \boldsymbol{x}_k^\natural \\ \hline \widetilde{\boldsymbol{h}}_k^t - \boldsymbol{h}_k^\natural \\ \hline \widetilde{\boldsymbol{x}}_k^t - \boldsymbol{x}_k^\natural \end{bmatrix} + \begin{bmatrix} \mathscr{A}_k(\boldsymbol{e})\boldsymbol{x}_k^t \\ \mathscr{A}_k^*(\boldsymbol{e})\boldsymbol{h}_k^t \\ \hline \mathscr{A}_k(\boldsymbol{e})\boldsymbol{x}_k^t \\ \hline \mathscr{A}_k^*(\boldsymbol{e})\boldsymbol{h}_k^t \end{bmatrix}, \tag{8.26}$$

where $\boldsymbol{H}_k = \int_0^1 \nabla_z^2 f_{\text{clean}}(z(\tau))\, d\tau$ with $z(\tau) := z^\natural + \tau\left(\widetilde{z}^t - z^\natural\right)$ and $\mathscr{A}_k(\boldsymbol{e}) = \sum_{j=1}^m e_j \boldsymbol{b}_j \boldsymbol{a}_{kj}^*$ and $\mathscr{A}_k^*(\boldsymbol{e}) = \sum_{j=1}^m \overline{e_j} \boldsymbol{a}_{kj} \boldsymbol{b}_j^*$. Since $z(\tau)$ lies between $\widetilde{z}^t$ and $z^\natural$, we derive from the assumption (8.20) that for all $\tau \in [0, 1]$,

$$\operatorname{dist}(z(\tau), z^\natural) \le \xi \le \delta,$$

$$\max_{1\le i\le s, 1\le j\le m} \left|\boldsymbol{a}_{ij}^*\left(\boldsymbol{x}_i(\tau) - \boldsymbol{x}_i^\natural\right)\right| \le \frac{C_3}{\sqrt{s}\log^{3/2} m},$$

$$\max_{1\le i\le s, 1\le j\le m} \left|\boldsymbol{b}_j^* \boldsymbol{h}_i(\tau)\right| \le \frac{C_4\mu}{\sqrt{m}} \log^2 m,$$

for some constants $C_3, C_4 > 0$ and the constant $\xi > 0$ being sufficiently small.

For simplicity, denote $\widehat{z}_k^{t+1} = [\widehat{\boldsymbol{h}}_k^{t+1*}\ \widehat{\boldsymbol{x}}_k^{t+1*}]^*$. Substituting (8.26) to (8.22), one has

$$\begin{bmatrix} \widehat{z}_k^{t+1} - z_k^\natural \\ \hline \widehat{z}_k^{t+1} - z_k^\natural \end{bmatrix} = \boldsymbol{\varphi}_k^t + \boldsymbol{\psi}_k^t, \tag{8.27}$$

where

$$\boldsymbol{\varphi}_k^t = (\boldsymbol{I} - \eta \boldsymbol{W}_k \boldsymbol{H}_k) \begin{bmatrix} \widetilde{z}_k^t - z_k^\natural \\ \hline \widetilde{z}_k^t - z_k^\natural \end{bmatrix}, \quad \boldsymbol{\psi}_k^t = \begin{bmatrix} \mathscr{A}_k(\boldsymbol{e})\boldsymbol{x}_k^t \\ \mathscr{A}_k^*(\boldsymbol{e})\boldsymbol{h}_k^t \\ \hline \mathscr{A}_k(\boldsymbol{e})\boldsymbol{x}_k^t \\ \hline \mathscr{A}_k^*(\boldsymbol{e})\boldsymbol{h}_k^t \end{bmatrix}.$$

Take the Euclidean norm of both sides of (8.27) to arrive

$$\|\boldsymbol{\varphi}_k^t + \boldsymbol{\psi}_k^t\|_2 \le \|\boldsymbol{\varphi}_k^t\|_2 + \|\boldsymbol{\psi}_k^t\|_2. \tag{8.28}$$

We first control the second Euclidean norm at the right-hand side of Eq. (8.28):

$$\|\boldsymbol{\psi}_k^t\|_2^2 = 2\left(\|\mathscr{A}_k(\boldsymbol{e})\|^2 \|\boldsymbol{x}_k^t\|_2^2 + \|\mathscr{A}_k^*(\boldsymbol{e})\|^2 \|\boldsymbol{h}_k^t\|_2^2\right) \leq 16 \|\mathscr{A}_k(\boldsymbol{e})\|^2, \tag{8.29}$$

where we use the fact that $\max\{\|\boldsymbol{x}_k\|_2, \|\boldsymbol{h}_k\|_2\} \leq 2$ for $1 \leq k \leq s$. Based on the paper [13, Section C.2], the squared Euclidean norm of $\boldsymbol{\varphi}_k^t$ is bounded by

$$\|\boldsymbol{\varphi}_k^t\|_2^2 \leq 2(1-\eta/8)\left\|\widetilde{\boldsymbol{z}}_k^t - \boldsymbol{z}_k^\natural\right\|_2^2, \tag{8.30}$$

under the assumption (8.20). We thus conclude that

$$\|\boldsymbol{\varphi}_k^t + \boldsymbol{\psi}_k^t\|_2 \leq \sqrt{2}(1-\eta/8)^{1/2}\left\|\widetilde{\boldsymbol{z}}_k^t - \boldsymbol{z}_k^\natural\right\|_2 + 4\|\mathscr{A}_k(\boldsymbol{e})\|, \tag{8.31}$$

and hence

$$\begin{aligned}\left\|\widetilde{\boldsymbol{z}}_k^{t+1} - \boldsymbol{z}_k^\natural\right\|_2 \leq \left\|\widehat{\boldsymbol{z}}_k^{t+1} - \boldsymbol{z}_k^\natural\right\|_2 &\leq \sqrt{2}/2\,\|\boldsymbol{\varphi}_k^t + \boldsymbol{\psi}_k^t\|_2 \\ &\leq (1-\eta/16)\left\|\widetilde{\boldsymbol{z}}_k^t - \boldsymbol{z}_k^\natural\right\|_2 + 3\|\mathscr{A}_k(\boldsymbol{e})\|.\end{aligned} \tag{8.32}$$

Integrate the above inequality (8.32) for $i = 1, \ldots, s$, we further get

$$\operatorname{dist}\left(\boldsymbol{z}^{t+1}, \boldsymbol{z}^\natural\right) \leq (1-\eta/16)\operatorname{dist}\left(\boldsymbol{z}^t, \boldsymbol{z}^\natural\right) + 3\sqrt{s}\max_{1\leq k\leq s}\|\mathscr{A}_k(\boldsymbol{e})\|. \tag{8.33}$$

Lemma 8.3 *Suppose the induction hypotheses hold true for t-th iteration and the number of measurements obeys $m \gg (\mu^2+\sigma^2)s^2K\log^8 m$. Then with probability at least $1 - O(m^{-9})$,*

$$\max_{1\leq i\leq s, 1\leq j\leq m}\left|\boldsymbol{b}_l^*\widetilde{\boldsymbol{h}}_i^{t+1}\right| \cdot \|\boldsymbol{h}_i^\natural\|_2^{-1} \leq C_4\frac{\mu}{\sqrt{m}}\log^2 m, \tag{8.34}$$

provided that C_4 is sufficiently large and the step size obeys $\eta > 0$ and $\eta \asymp s^{-1}$.

Proof Similar to the strategy used in [13, Section C.4], it suffices to control $|\boldsymbol{b}_l^*\frac{1}{\alpha_i^t}\boldsymbol{h}_i^{t+1}|$ to finish the proof, since

$$\begin{aligned}\max_{1\leq i\leq s, 1\leq l\leq m}\left|\boldsymbol{b}_l^*\frac{1}{\alpha_i^{t+1}}\boldsymbol{h}_i^{t+1}\right| &\leq \left|\frac{\alpha_i^t}{\alpha_i^{t+1}}\right|\max_{1\leq i\leq s, 1\leq l\leq m}\left|\boldsymbol{b}_l^*\frac{1}{\alpha_i^t}\boldsymbol{h}_i^{t+1}\right| \\ &\leq (1+\delta)\left|\boldsymbol{b}_l^*\frac{1}{\alpha_i^t}\boldsymbol{h}_i^{t+1}\right|\end{aligned} \tag{8.35}$$

for some small $\delta \asymp 1/\log^2 m$, where the last step bases on

$$\left|\frac{\alpha_i^{t+1}}{\alpha_i^t} - 1\right| \lesssim \frac{1}{\log^2 m} \leq \delta. \tag{8.36}$$

The gradient update rule for $\boldsymbol{h}_i^{t+1}$ is written as

$$\frac{1}{\alpha_i^t}\boldsymbol{h}_i^{t+1} = \widetilde{\boldsymbol{h}}_i^t - \eta\xi_i \sum_{j=1}^{m}\sum_{k=1}^{s} \boldsymbol{b}_j\boldsymbol{b}_j^*\left(\widetilde{\boldsymbol{h}}_k^t\widetilde{\boldsymbol{x}}_k^{t*} - \boldsymbol{h}_k^\natural\boldsymbol{h}_k^{\natural*}\right)\boldsymbol{a}_{kj}\boldsymbol{a}_{ij}^*\widetilde{\boldsymbol{x}}_i^t + \eta\xi_i\sum_{j=1}^{m} e_j\boldsymbol{b}_j\boldsymbol{a}_{ij}^*\widetilde{\boldsymbol{x}}_i^t, \tag{8.37}$$

where $\xi_i = \frac{1}{\|\widetilde{\boldsymbol{x}}_i^t\|_2^2}$ and $\widetilde{\boldsymbol{h}}_i^t = \frac{1}{\alpha_i^t}\boldsymbol{h}_i^t$ and $\widetilde{\boldsymbol{x}}_i^t = \alpha_i^t\boldsymbol{x}_i^t$ for $i = 1, \ldots, s$. The formula (8.37) can be further decomposed into the following terms:

$$\begin{aligned}\frac{1}{\alpha_i^t}\boldsymbol{h}_i^{t+1} &= \widetilde{\boldsymbol{h}}_i^t - \eta\xi_i \sum_{j=1}^{m}\sum_{k=1}^{s} \boldsymbol{b}_j\boldsymbol{b}_j^*\widetilde{\boldsymbol{h}}_k^t\widetilde{\boldsymbol{x}}_k^{t*}\boldsymbol{a}_{kj}\boldsymbol{a}_{ij}^*\widetilde{\boldsymbol{x}}_i^t + \eta\xi_i \sum_{j=1}^{m}\sum_{k=1}^{s} \boldsymbol{b}_j\boldsymbol{b}_j^*\boldsymbol{h}_k^\natural\boldsymbol{x}_k^{\natural*}\boldsymbol{a}_{kj}\boldsymbol{a}_{ij}^*\widetilde{\boldsymbol{x}}_i^t \\ &\quad + \eta\xi_i\sum_{j=1}^{m} e_j\boldsymbol{b}_j\boldsymbol{a}_{ij}^*\widetilde{\boldsymbol{x}}_i^t \\ &= \widetilde{\boldsymbol{h}}_i^t - \eta\xi_i\sum_{k=1}^{s}\widetilde{\boldsymbol{h}}_k^t\left\|\boldsymbol{x}_k^\natural\right\|_2^2 - \eta\xi_i\boldsymbol{v}_{i1} - \eta\xi_i\boldsymbol{v}_{i2} + \eta\xi_i\boldsymbol{v}_{i3} + \eta\xi_i\boldsymbol{v}_{i4},\end{aligned} \tag{8.38}$$

where

$$\boldsymbol{v}_{i1} = \sum_{j=1}^{m}\sum_{k=1}^{s}\boldsymbol{b}_j\boldsymbol{b}_j^*\widetilde{\boldsymbol{h}}_k^t\left(\widetilde{\boldsymbol{x}}_k^{t*}\boldsymbol{a}_{kj}\boldsymbol{a}_{ij}^*\widetilde{\boldsymbol{x}}_i^t - \boldsymbol{x}_k^{\natural*}\boldsymbol{a}_{kj}\boldsymbol{a}_{ij}^*\boldsymbol{x}_i^\natural\right)$$

$$\boldsymbol{v}_{i2} = \sum_{j=1}^{m}\sum_{k=1}^{s}\boldsymbol{b}_j\boldsymbol{b}_j^*\widetilde{\boldsymbol{h}}_k^t\left(\boldsymbol{x}_k^{\natural*}\boldsymbol{a}_{kj}\boldsymbol{a}_{ij}^*\boldsymbol{x}_i^\natural - \|\boldsymbol{x}_k^\natural\|_2^2\right)$$

$$\boldsymbol{v}_{i3} = \sum_{j=1}^{m}\sum_{k=1}^{s}\boldsymbol{b}_j\boldsymbol{b}_j^*\boldsymbol{h}_k^\natural\boldsymbol{x}_k^{\natural*}\boldsymbol{a}_{kj}\boldsymbol{a}_{ij}^*\widetilde{\boldsymbol{x}}_i^t$$

$$\boldsymbol{v}_{i4} = \sum_{j=1}^{m} e_j\boldsymbol{b}_j\boldsymbol{a}_{ij}^*\widetilde{\boldsymbol{x}}_i^t,$$

which bases on the fact that $\sum_{j=1}^{m}\boldsymbol{b}_j\boldsymbol{b}_j^* = \boldsymbol{I}_K$. In what follows, we bound the above four terms, respectively.

1. We start with $|\boldsymbol{b}_l^* \boldsymbol{v}_{i1}|$ via the operation that

$$|\boldsymbol{b}_l^* \boldsymbol{v}_{i1}| = \left| \sum_{j=1}^{m} \boldsymbol{b}_l^* \boldsymbol{b}_j \boldsymbol{b}_j^* \left[\sum_{k=1}^{s} \widetilde{\boldsymbol{h}}_k^t \left(\boldsymbol{a}_{ij}^* \left(\widetilde{\boldsymbol{x}}_i^t - \boldsymbol{x}_i^\natural \right) \left(\boldsymbol{a}_{kj}^* \widetilde{\boldsymbol{x}}_k^t \right)^* + \boldsymbol{a}_{ij}^* \boldsymbol{x}_i^\natural \left(\boldsymbol{a}_{kj}^* \left(\widetilde{\boldsymbol{x}}_k^t - \boldsymbol{x}_k^\natural \right) \right)^* \right) \right] \right| \leq s \sum_{j=1}^{m} |\boldsymbol{b}_l^* \boldsymbol{b}_j| \left\{ \max_{1 \leq k \leq s, 1 \leq j \leq m} \left| \boldsymbol{b}_j^* \widetilde{\boldsymbol{h}}_k^t \right| \right\} \cdot \left\{ \max_{1 \leq k \leq s, 1 \leq j \leq m} \left| \boldsymbol{a}_{kj}^* (\widetilde{\boldsymbol{x}}_k^t - \boldsymbol{x}_k^\natural) \right| \left(\left| \boldsymbol{a}_{kj}^* \widetilde{\boldsymbol{x}}_k^t \right| + \left| \boldsymbol{a}_{kj}^* \boldsymbol{x}_k^\natural \right| \right) \right\}. \tag{8.39}$$

Based on the inductive hypothesis (8.17c) and the concentration inequality [13]

$$\max_{1 \leq i \leq s, 1 \leq j \leq m} \left| \boldsymbol{a}_{ij}^* \boldsymbol{x}_i^\natural \right| \leq 5\sqrt{\log m}, \tag{8.40}$$

with probability at least $1 - O(m^{-10})$, it yields

$$\max_{1 \leq k \leq s, 1 \leq j \leq m} \left| \boldsymbol{a}_{kj}^* \widetilde{\boldsymbol{x}}_k^t \right| \leq \max_{1 \leq k \leq s, 1 \leq j \leq m} \left| \boldsymbol{a}_{kj}^* \left(\widetilde{\boldsymbol{x}}_k^t - \boldsymbol{x}_k^\natural \right) \right| + \max_{1 \leq k \leq s, 1 \leq j \leq m} \left| \boldsymbol{a}_{kj}^* \boldsymbol{x}_k^\natural \right| \leq 6\sqrt{\log m}, \tag{8.41}$$

as long as m is sufficiently large. We further derive that

$$\max_{1 \leq k \leq s, 1 \leq j \leq m} \left| \boldsymbol{a}_{kj}^* \left(\widetilde{\boldsymbol{x}}_k^t - \boldsymbol{x}_k^\natural \right) \right| \left(\left| \boldsymbol{a}_{kj}^* \widetilde{\boldsymbol{x}}_k^t \right| + \left| \boldsymbol{a}_{kj}^* \widetilde{\boldsymbol{x}}_k^\natural \right| \right) \leq \frac{1}{\sqrt{s} \log^{3/2} m} \cdot 11\sqrt{\log m} \leq 11 C_3 \frac{1}{\log m}. \tag{8.42}$$

Substituting (8.42) into (8.39) and combining lemma [13, Lemma 48] such that

$$\sum_{j=1}^{m} |\boldsymbol{b}_l^* \boldsymbol{b}_j| \leq 4 \log m, \tag{8.43}$$

we get

$$\begin{aligned} |\boldsymbol{b}_l^* \boldsymbol{v}_{i1}| &\lesssim s \log m \cdot \left\{ \max_{1 \leq k \leq s, 1 \leq j \leq m} \left| \boldsymbol{b}_j^* \widetilde{\boldsymbol{h}}_k^t \right| \right\} \cdot C_3 \frac{1}{\log m} \\ &\lesssim s C_3 \max_{1 \leq k \leq s, 1 \leq j \leq m} \left| \boldsymbol{b}_j^* \widetilde{\boldsymbol{h}}_k^t \right| \\ &\leq 0.1 s \max_{1 \leq k \leq s, 1 \leq j \leq m} \left| \boldsymbol{b}_j^* \widetilde{\boldsymbol{h}}_k^t \right|, \end{aligned} \tag{8.44}$$

as long as C_3 is sufficiently small.

2. Regarding to $|\boldsymbol{b}_l^* \boldsymbol{v}_{i3}|$, one has

$$|\boldsymbol{b}_l^* \boldsymbol{v}_{i3}| \leq \left| \sum_{j=1}^{m} \boldsymbol{b}_l^* \boldsymbol{b}_j \boldsymbol{b}_j^* \left(\sum_{k=1}^{s} \boldsymbol{h}_k^{\natural} \boldsymbol{x}_k^{\natural *} \boldsymbol{a}_{kj} \right) \boldsymbol{a}_{ij}^* \boldsymbol{x}_i^{\natural} \right| + \left| \sum_{j=1}^{m} \boldsymbol{b}_l^* \boldsymbol{b}_j \boldsymbol{b}_j^* \left(\sum_{k=1}^{s} \boldsymbol{h}_k^{\natural} \boldsymbol{x}_k^{\natural *} \boldsymbol{a}_{kj} \right) \boldsymbol{a}_{ij}^* \left(\widetilde{\boldsymbol{x}}_i^t - \boldsymbol{x}_i^{\natural} \right) \right|. \tag{8.45}$$

Lemma 8.4 *Suppose $m \gg s^2 K \log m$ for some sufficiently large constant $C > 0$. Then with probability at least $1 - O(m^{-10})$, there is*

$$\left| \sum_{j=1}^{m} \boldsymbol{b}_l^* \boldsymbol{b}_j \boldsymbol{b}_j^* \left(\sum_{k=1}^{s} \boldsymbol{h}_k^{\natural} \boldsymbol{x}_k^{\natural *} \boldsymbol{a}_{kj} \right) \boldsymbol{a}_{ij}^* \boldsymbol{x}_i^{\natural} - \boldsymbol{b}_l^* \boldsymbol{h}_i^{\natural} \right| \lesssim \frac{\mu}{\sqrt{m}}. \tag{8.46}$$

Proof See Appendix 8.3.1.

Regarding to the second term in (8.45), we exploit the same technical method as in controlling $|\boldsymbol{b}_l^* \boldsymbol{v}_{i1}|$, which yields

$$\begin{aligned}
&\left| \sum_{j=1}^{m} \boldsymbol{b}_l^* \boldsymbol{b}_j \boldsymbol{b}_j^* \left(\sum_{k=1}^{s} \boldsymbol{h}_k^{\natural} \boldsymbol{x}_k^{\natural *} \boldsymbol{a}_{kj} \right) \boldsymbol{a}_{ij}^* \left(\widetilde{\boldsymbol{x}}_i^t - \boldsymbol{x}_i^{\natural} \right) \right| \\
&\leq s \sum_{j=1}^{m} |\boldsymbol{b}_l^* \boldsymbol{b}_j| \left\{ \max_{1 \leq k \leq s, 1 \leq j \leq m} \left| \boldsymbol{b}_j^* \boldsymbol{h}_k^{\natural} \right| \right\} \cdot \left\{ \max_{1 \leq k \leq s, 1 \leq j \leq m} \left| \boldsymbol{a}_{kj}^* \left(\widetilde{\boldsymbol{x}}_k^t - \boldsymbol{x}_k^{\natural} \right) \right| \right\} \cdot \\
&\quad \left\{ \max_{1 \leq k \leq s, 1 \leq j \leq m} \left| \boldsymbol{a}_{kj}^* \boldsymbol{x}_k^{\natural} \right| \right\} \\
&\leq 4s \log m \cdot \frac{\mu}{\sqrt{m}} \cdot C_3 \frac{1}{\sqrt{s} \log^{3/2} m} \cdot 5\sqrt{\log m} \\
&\lesssim C_3 \frac{\sqrt{s}\mu}{\sqrt{m}},
\end{aligned} \tag{8.47}$$

where the second step arises from the incoherence, the induction hypothesis (8.17c) and the condition (8.40) and [13, Lemma 48]. Combining the above inequalities and the incoherence, one achieves

$$|\boldsymbol{b}_l^* \boldsymbol{v}_{i3}| \lesssim \left| \boldsymbol{b}_l^* \boldsymbol{h}_i^{\natural} \right| + \frac{\mu}{\sqrt{m}} + C_3 \frac{\sqrt{s}\mu}{\sqrt{m}} \lesssim (1 + C_3\sqrt{s}) \frac{\mu}{\sqrt{m}}, \tag{8.48}$$

as long as picking up sufficiently small $C_3 > 0$.

3. We further move to control $|\boldsymbol{b}_l^* \boldsymbol{v}_{i2}|$. The idea of proof is based on the strategy in [13, Section C.4], which groups $\{\boldsymbol{b}_j\}_{1 \le j \le m}$ into bins each containing τ adjacent vectors. Similarly to the paper [13], we assume m/τ to be an integer. For $0 \le l \le m - \tau$, one has

$$\begin{aligned}\boldsymbol{b}_1^* \sum_{j=1}^{\tau} \boldsymbol{b}_{l+j} \boldsymbol{b}_{l+j}^* \Big(\sum_{k=1}^{s} \widetilde{\boldsymbol{h}}_k^t z_{ijkl} \Big) &= \boldsymbol{b}_1^* \sum_{j=1}^{\tau} \boldsymbol{b}_{l+1} \boldsymbol{b}_{l+1}^* \Big(\sum_{k=1}^{s} \widetilde{\boldsymbol{h}}_k^t z_{ijkl} \Big) \\ &\quad + \boldsymbol{b}_1^* \sum_{j=1}^{\tau} \Big(\boldsymbol{b}_{l+j} \boldsymbol{b}_{l+j}^* - \boldsymbol{b}_{l+1} \boldsymbol{b}_{l+1}^* \Big) \Big(\sum_{k=1}^{s} \widetilde{\boldsymbol{h}}_k^t z_{ijkl} \Big) \\ &= p_{i\tau 1} + p_{i\tau 2} + p_{i\tau 3},\end{aligned} \tag{8.49}$$

where

$$\begin{aligned} z_{ijkl} &= \boldsymbol{x}_k^{\natural *} \boldsymbol{a}_{k,j+l} \boldsymbol{a}_{i,j+l}^* \boldsymbol{x}_i^{\natural} - \left\| \boldsymbol{x}_i^{\natural} \right\|_2^2, \\ p_{i\tau 1} &= \sum_{k=1}^{s} \Big(\sum_{j=1}^{\tau} z_{ijkl} \Big) \boldsymbol{b}_1^* \boldsymbol{b}_{l+1} \boldsymbol{b}_{l+1}^* \widetilde{\boldsymbol{h}}_k^t, \\ p_{i\tau 2} &= \boldsymbol{b}_1^* \sum_{j=1}^{\tau} \left(\boldsymbol{b}_{l+j} - \boldsymbol{b}_{l+1} \right) \boldsymbol{b}_{l+j}^* \sum_{k=1}^{s} \widetilde{\boldsymbol{h}}_k^t z_{ijkl}, \\ p_{i\tau 3} &= \boldsymbol{b}_1^* \sum_{j=1}^{\tau} \boldsymbol{b}_{l+1} (\boldsymbol{b}_{l+j} - \boldsymbol{b}_{l+1})^* \sum_{k=1}^{s} \widetilde{\boldsymbol{h}}_k^t z_{ijkl}. \end{aligned}$$

We will control three terms in (8.49), respectively.

(a) According to [13, Section C.4], with probability at least $1 - O(m^{-10})$,

$$\left| \sum_{j=1}^{\tau} z_{ijkl} \right| \le \left| \sum_{j=1}^{\tau} \left(\max \left\{ \left| \boldsymbol{a}_{k,l+j}^* \boldsymbol{x}_k^{\natural} \right|^2, \left| \boldsymbol{a}_{i,l+j}^* \boldsymbol{x}_i^{\natural} \right|^2 \right\} - \left\| \boldsymbol{x}_i^{\natural} \right\|_2^2 \right) \right| \lesssim \sqrt{\tau \log m}. \tag{8.50}$$

Combining above bound, we control the first term in (8.49) as

$$|p_{i\tau 1}| \lesssim s \sqrt{\tau \log m} |\boldsymbol{b}_1^* \boldsymbol{b}_{l+1}| \max_{1 \le k \le s, 1 \le j \le m} \left| \boldsymbol{b}_l^* \widetilde{\boldsymbol{h}}_k^t \right|. \tag{8.51}$$

The summation over all bins is given as

$$\sum_{d=0}^{\frac{m}{\tau}-1}\left|\sum_{j=1}^{\tau} z_{ijk(d\tau)}\boldsymbol{b}_1^*\boldsymbol{b}_{d\tau+1}\boldsymbol{b}_{d\tau+1}^*\widetilde{\boldsymbol{h}}_k^t\right|$$

$$\lesssim s\sqrt{\tau\log m}\sum_{d=0}^{\frac{m}{r}-1}\left|\boldsymbol{b}_1^*\boldsymbol{b}_{d\tau+1}\right|\max_{1\le k\le s,1\le j\le m}\left|\boldsymbol{b}_l^*\widetilde{\boldsymbol{h}}_k^t\right|. \tag{8.52}$$

Substituting the bound

$$\sum_{d=0}^{\frac{m}{\tau}-1}\left|\boldsymbol{b}_1^*\boldsymbol{b}_{d\tau+1}\right|\le\frac{K}{m}+O\left(\frac{\log m}{\tau}\right), \tag{8.53}$$

provided in [13, Section C.4], into the inequality (8.52) yields

$$\sum_{d=0}^{\frac{m}{\tau}-1}\left|\sum_{j=1}^{\tau} z_{ijk(d\tau)}\boldsymbol{b}_1^*\boldsymbol{b}_{d\tau+1}\boldsymbol{b}_{d\tau+1}^*\widetilde{\boldsymbol{h}}_k^t\right|$$

$$\lesssim\left(\frac{sK\sqrt{\tau\log m}}{m}+\sqrt{\frac{s^2\log^3 m}{\tau}}\right)\max_{1\le k\le s,1\le j\le m}\left|\boldsymbol{b}_l^*\widetilde{\boldsymbol{h}}_k^t\right|$$

$$\le 0.1\max_{1\le k\le s,1\le j\le m}\left|\boldsymbol{b}_l^*\widetilde{\boldsymbol{h}}_k^t\right|, \tag{8.54}$$

as long as $m\gg Ks\sqrt{\tau\log m}$ and $\tau\gg s^2\log^3 m$.

(b) The second term of (8.49), $p_{i\tau 2}$, is controlled by

$$|p_{i\tau 2}|\le\max_{1\le k\le s,1\le l\le m}\left|\boldsymbol{b}_l^*\widetilde{\boldsymbol{h}}_k^t\right|\sqrt{\sum_{j=1}^{\tau}\left|\boldsymbol{b}_1^*(\boldsymbol{b}_{l+j}-\boldsymbol{b}_{l+1})\right|^2}\cdot$$

$$\sqrt{\sum_{j=1}^{\tau}\sum_{k=1}^{s}\left(|\max\left\{\left|\boldsymbol{a}_{k,l+j}^*\boldsymbol{x}_k^\natural\right|^2,\left|\boldsymbol{a}_{i,l+j}^*\boldsymbol{x}_i^\natural\right|^2\right\}-\left\|\boldsymbol{x}_k^\natural\right\|_2^2\right)^2}$$

$$\lesssim\sqrt{s\tau}\max_{1\le k\le s,1\le l\le m}\left|\boldsymbol{b}_l^*\widetilde{\boldsymbol{h}}_k^t\right|\sqrt{\sum_{i=1}^{\tau}\left|\boldsymbol{b}_1^*(\boldsymbol{b}_{l+j}-\boldsymbol{b}_{l+1})\right|^2}, \tag{8.55}$$

where the first inequality is due to Cauchy–Schwarz and the second step holds because of the following lemma.

Lemma 8.5 *Suppose* $\tau \geq C \log^4 m$ *for some sufficiently large constant* $C > 0$, *with probability at least* $1 - O(m^{-10})$, *one has*

$$\sum_{j=1}^{\tau} \sum_{k=1}^{s} \left(| \max \left\{ \left| \boldsymbol{a}_{k,l+j}^* \boldsymbol{x}_k^\natural \right|^2, \left| \boldsymbol{a}_{i,l+j}^* \boldsymbol{x}_i^\natural \right|^2 \right\} - \left\| \boldsymbol{x}_k^\natural \right\|_2^2 \right)^2 \lesssim s\tau. \tag{8.56}$$

Proof This claim can be identified easily from [13, Appendix D.3.1].

We further sum over all bins of size τ to obtain

$$\begin{aligned} &\left| \boldsymbol{b}_1^* \sum_{d=0}^{\frac{m}{\tau}-1} \sum_{j=1}^{\tau} \left(\boldsymbol{b}_{d\tau+j} - \boldsymbol{b}_{d\tau+1} \right) \boldsymbol{b}_{d\tau+j}^* \sum_{k=1}^{s} \widetilde{\boldsymbol{h}}_k^t z_{ijk(d\tau)} \right| \\ &\leq \left\{ \sqrt{s\tau} \sum_{d=0}^{\frac{m}{\tau}-1} \sqrt{\sum_{i=1}^{\tau} \left| \boldsymbol{b}_1^* (\boldsymbol{b}_{d\tau+j} - \boldsymbol{b}_{d\tau+1}) \right|^2} \right\} \cdot \max_{1 \leq k \leq s, 1 \leq l \leq m} \left| \boldsymbol{b}_l^* \widetilde{\boldsymbol{h}}_k^t \right| \\ &\leq 0.1 \sqrt{s} \max_{1 \leq k \leq s, 1 \leq l \leq m} \left| \boldsymbol{b}_l^* \widetilde{\boldsymbol{h}}_k^t \right|. \end{aligned} \tag{8.57}$$

Here, the last line arises from [13, Lemma 51] such that for any small constant $c > 0$,

$$\sum_{d=0}^{\frac{m}{\tau}-1} \sqrt{\sum_{j=1}^{\tau} \left| \boldsymbol{b}_1^* (\boldsymbol{b}_{d\tau+j} - \boldsymbol{b}_{d\tau+1}) \right|^2} \leq c \frac{1}{\sqrt{\tau}}, \tag{8.58}$$

as long as $m \gg \tau K \log m$.

(c) The third term of (8.49), $p_{i\tau 3}$, obeys that

$$\begin{aligned} |p_{i\tau 3}| \leq & |\boldsymbol{b}_1^* \boldsymbol{b}_{l+1}| \left\{ \sum_{j=1}^{\tau} \sum_{k=1}^{s} \left(| \max \left\{ \left| \boldsymbol{a}_{k,l+j}^* \boldsymbol{x}_k^\natural \right|^2, \left| \boldsymbol{a}_{i,l+j}^* \boldsymbol{x}_i^\natural \right|^2 \right\} - \left\| \boldsymbol{x}_k^\natural \right\|_2^2 \right)^2 \right\} \cdot \\ & \max_{1 \leq k \leq s, 1 \leq l \leq m-\tau, 1 \leq j \leq \tau} \left| (\boldsymbol{b}_{l+j} - \boldsymbol{b}_{l+1})^* \widetilde{\boldsymbol{h}}_k^t \right| \\ \lesssim & s\tau |\boldsymbol{b}_1^* \boldsymbol{b}_{l+1}| \max_{1 \leq k \leq s, 1 \leq l \leq m-\tau, 1 \leq j \leq \tau} \left| (\boldsymbol{b}_{l+j} - \boldsymbol{b}_{l+1})^* \widetilde{\boldsymbol{h}}_k^t \right|, \end{aligned} \tag{8.59}$$

where the last line relies on the inequality (8.56) and the Cauchy–Schwarz inequality.

The summation over all bins is given as

$$\begin{aligned}
&\sum_{d=0}^{\frac{m}{\tau}-1}\left|\boldsymbol{b}_1^*\sum_{j=1}^{\tau}\boldsymbol{b}_{d\tau+1}(\boldsymbol{b}_{d\tau+j}-\boldsymbol{b}_{d\tau+1})^*\sum_{k=1}^{s}\widetilde{\boldsymbol{h}}_k^t z_{ijk(d\tau)}\right| \\
&\lesssim \tau s\sum_{d=0}^{\frac{m}{\tau}-1}|\boldsymbol{b}_1^*\boldsymbol{b}_{d\tau+1}|\cdot\max_{1\le k\le s,1\le l\le m-\tau,1\le j\le\tau}\left|(\boldsymbol{b}_{l+j}-\boldsymbol{b}_{l+1})^*\widetilde{\boldsymbol{h}}_k^t\right| \\
&\lesssim s\log m\max_{1\le k\le s,1\le l\le m-\tau,1\le j\le\tau}\left|(\boldsymbol{b}_{l+j}-\boldsymbol{b}_{l+1})^*\widetilde{\boldsymbol{h}}_k^t\right| \\
&\lesssim cC_4\frac{s\mu}{\sqrt{m}}\log^2 m, \qquad (8.60)
\end{aligned}$$

where the last relation makes use of (8.53) and the claim

$$\max_{1\le j\le\tau}|(\boldsymbol{b}_j-\boldsymbol{b}_1)^*\widetilde{\boldsymbol{h}}_k^t|\le cC_4\frac{\mu}{\sqrt{m}}\log m, \qquad (8.61)$$

for some sufficiently small constant $c>0$, provided that $m\gg\tau K\log^4 m$.

(d) Combining the above results together, we get

$$|\boldsymbol{b}_1^*\boldsymbol{v}_{i2}|\le(0.1+0.1\sqrt{s})\max_{1\le k\le s,1\le l\le m}\left|\boldsymbol{b}_l^*\widetilde{\boldsymbol{h}}_k^\top\right|+O\left(cC_4\frac{s\mu}{\sqrt{m}}\log^2 m\right). \qquad (8.62)$$

4. We end the proof with controlling $|\boldsymbol{b}_l^*\boldsymbol{v}_{i4}|$:

$$\begin{aligned}
|\boldsymbol{b}_l^*\boldsymbol{v}_{i4}|&=\left|\sum_{j=1}^{m}\boldsymbol{b}_l^*\boldsymbol{b}_j e_j\boldsymbol{a}_{ij}^*\widetilde{\boldsymbol{x}}_i^t\right|\le\sum_{j=1}^{m}|\boldsymbol{b}_l^*\boldsymbol{b}_j|\cdot\left\{\max_{1\le k\le s,1\le j\le m}\left|\boldsymbol{a}_{kj}^*\widetilde{\boldsymbol{x}}_k^t\right|\right\}\cdot|e_j| \\
&\le 4\log m\cdot 6\sqrt{\log m}\cdot\frac{\sigma^2}{m}, \qquad (8.63)
\end{aligned}$$

where the last step arises from the inequality (8.41), (8.43) and the assumption $|e_j|\lesssim\sigma^2/m\ll 1$. It thus yields

$$|\boldsymbol{b}_l^*\boldsymbol{v}_{i4}|\lesssim\sigma^2\frac{\log^{3/2} m}{m}\le\log m, \qquad (8.64)$$

as long as $m\gg\sigma^2\sqrt{\log m}$.

5. Putting the above results together, there exists some constant $C_8 > 0$ such that

$$\begin{aligned}
\left|\boldsymbol{b}_l^* \widetilde{\boldsymbol{h}}_i^{t+1}\right| &\le (1+\delta)\Bigg\{ \left|\boldsymbol{b}_l^* \widetilde{\boldsymbol{h}}_i^t\right| - \eta\xi_i \sum_{k=1}^{s} \left|\boldsymbol{b}_l^* \widetilde{\boldsymbol{h}}_k^t\right| + (1 + 0.1\sqrt{s} + 0.1s) \\
&\quad \max_{1\le k\le s, 1\le j\le m} \left|\boldsymbol{b}_j^* \widetilde{\boldsymbol{h}}_k^t\right| + C_8(1 + C_3\sqrt{s})\eta\xi_i \frac{\mu}{\sqrt{m}} \\
&\quad + C_8 c C_4 \eta\xi_i \frac{s\mu}{\sqrt{m}} \log^2 m + C_8\eta\xi_i \log m \Bigg\} \\
&\overset{\text{(i)}}{\le} \left(1 + O\left(\frac{1}{\log^2 m}\right)\right) \Bigg\{ (1 - 0.7s\eta\xi_i) C_4 \frac{\mu}{\sqrt{m}} \log^2 m \\
&\quad + C_8(1 + C_3\sqrt{s})\eta\xi_i \frac{\mu}{\sqrt{m}} + C_8 c C_4 \eta\xi_i \frac{s\mu}{\sqrt{m}} \log^2 m + C_8\eta\xi_i \log m \Bigg\} \\
&\overset{\text{(ii)}}{\le} C_4 \frac{\mu}{\sqrt{m}} \log^2 m. \qquad (8.65)
\end{aligned}$$

Here, step (i) arises from the induction hypothesis (8.17d), step (ii) holds as log as $c > 0$ is sufficiently small, i.e., $(1+\delta)C_8\eta\xi_i c \gg 1$, and $\eta > 0$ is some sufficiently small constant, i.e., $\eta \asymp s^{-1}$. In order for the proof to go through, we need to pick the sample size that

$$m \gg (\mu^2 + \sigma^2)\tau K \log^4 m, \qquad (8.66)$$

where $\tau = c_{10} s^2 \log^4 m$ with some sufficiently large constant $c_{10} > 0$.

8.3.1 Proof of Lemma 8.4

Denote

$$w_{ij} = \boldsymbol{b}_l^* \boldsymbol{b}_j \boldsymbol{b}_j^* \left(\sum_{k=1}^{s} \boldsymbol{h}_k^\natural \boldsymbol{x}_k^{\natural *} \boldsymbol{a}_{kj} \right) \boldsymbol{a}_{ij}^* \boldsymbol{x}_i^\natural. \qquad (8.67)$$

Combining the fact that $\mathbb{E}[\boldsymbol{a}_{ij}\boldsymbol{a}_{ij}^*] = \boldsymbol{I}_K$, $\mathbb{E}[\boldsymbol{a}_{kj}\boldsymbol{a}_{ij}^*] = 0$ for $k \ne i$ and $\sum_{j=1}^m \boldsymbol{b}_j \boldsymbol{b}_j^* = \boldsymbol{I}_K$, we can represent the objective quantity as the sum of independent random variables,

$$\sum_{j=1}^{m} \boldsymbol{b}_l^* \boldsymbol{b}_j \boldsymbol{b}_j^* \left(\sum_{k=1}^{s} \boldsymbol{h}_k^\natural \boldsymbol{x}_k^{\natural *} \boldsymbol{a}_{kj} \right) \boldsymbol{a}_{ij}^* \boldsymbol{x}_i^\natural - \boldsymbol{b}_l^* \boldsymbol{h}_i^\natural = \sum_{j=1}^{m} \left(w_{ij} - \mathbb{E}(w_{ij}) \right). \qquad (8.68)$$

Based on the definition of sub-exponential norm [21], i.e., denoted as $\|\cdot\|_{\psi_1}$, we get

$$\begin{aligned}
\|w_{ij} - \mathbb{E}[w_{ij}]\|_{\psi_1} &\overset{\text{(i)}}{\leq} 2\|w_{ij}\|_{\psi_1} \\
&\overset{\text{(ii)}}{\leq} 4\sum_{k=1}^{s} \left|\boldsymbol{b}_l^* \boldsymbol{b}_j\right| \left|\boldsymbol{b}_j^* \boldsymbol{h}_k^\natural\right| \max_{1\leq q\leq s} \left\|\boldsymbol{a}_{qj}^* \boldsymbol{x}_q^\natural\right\|_{\psi_2}^2 \qquad (8.69) \\
&\overset{\text{(iii)}}{\lesssim} \left|\boldsymbol{b}_l^* \boldsymbol{b}_j\right| \frac{s\mu}{\sqrt{m}}, \qquad (8.70)
\end{aligned}$$

where (i) uses the centering property of the sub-exponential norm [21, Remark 5.18], (ii) arises from the relationship between the sub-exponential norm and the sub-Gaussian norm [21, Lemma 5.14], and (iii) occurs since the incoherence condition and the fact that $\|\boldsymbol{a}_{kj}^* \boldsymbol{x}_k^\natural\|_{\psi_2} \lesssim 1$. According to [13, Section C.4.1], one has

$$W := \sqrt{\frac{1}{m}\sum_{j=1}^{m} \|w_{ij} - \mathbb{E}[w_{ij}]\|_{\psi_1}^2} \asymp \frac{\mu}{\sqrt{m}} \frac{s\sqrt{K}}{m}. \qquad (8.71)$$

It can further invoke [10, Corollary 1] to obtain

$$\mathbb{P}\left(\left|\frac{1}{m}\sum_{j=1}^{m}(w_{ij} - \mathbb{E}[w_{ij}])\right| \geq t\right) \leq \exp\left(1 - \frac{m}{8}\min\left\{\frac{t}{2W}, \left(\frac{t}{2W}\right)^2\right\}\right). \qquad (8.72)$$

By taking $t = 2\varepsilon W$ for $\varepsilon \in (0, 1)$, we obtain with probability at least $1 - \exp(1 - m\varepsilon^2/8)$,

$$\sum_{j=1}^{m}(w_{ij} - \mathbb{E}[w_{ij}]) \leq 2\varepsilon W m \lesssim \varepsilon s\sqrt{K}\frac{\mu}{\sqrt{m}}. \qquad (8.73)$$

Thus, choosing $\varepsilon \asymp 1/s\sqrt{K}$, we conclude that with probability at least $1 - \exp(1 - \text{cm/s}^2\text{K})$ for some constant $c > 0$,

$$\left|\sum_{j=1}^{m}(w_{ij} - \mathbb{E}[w_{ij}])\right| \lesssim \frac{\mu}{\sqrt{m}}. \qquad (8.74)$$

We finished the proof by observing that $m \gg s^2 K \log m$ as claimed in the assumption.

Lemma 8.6 *With probability at least* $1-O(m^{-9})$*, there exists some constant* $C > 0$ *such that*

$$\min_{\alpha_i \in \mathbb{C}, |\alpha_i|=1} \left\{ \left\| \alpha_i \boldsymbol{h}_i^0 - \boldsymbol{h}_i^\natural \right\| + \left\| \alpha_i \boldsymbol{x}_i^0 - \boldsymbol{x}_i^\natural \right\| \right\} \leq \frac{\xi}{\kappa\sqrt{s}} \text{ and} \tag{8.75}$$

$$\min_{\alpha_i \in \mathbb{C}, |\alpha_i|=1} \left\{ \left\| \alpha_i \boldsymbol{h}_i^{0,(l)} - \boldsymbol{h}_i^\natural \right\| + \left\| \alpha_i \boldsymbol{x}_i^{0,(l)} - \boldsymbol{x}_i^\natural \right\| \right\} \leq \frac{\xi}{\kappa\sqrt{s}}, \tag{8.76}$$

and $||\alpha_i^0| - 1| < 1/4$*, for each* $1 \leq i \leq s$, $1 \leq l \leq m$*, provided that*

$$m \geq C(\mu^2 + \sigma^2) s\kappa^2 K \log m / \xi^2.$$

Lemma 8.7 *Suppose that* $m \gg (\mu^2 + \sigma^2) s^2 \kappa^2 K \log^3 m$*. Then with probability at least* $1 - O(m^{-9})$*,*

$$\operatorname{dist}\left(\boldsymbol{z}^{0,(l)}, \widetilde{\boldsymbol{z}}^0\right) \leq C_2 \frac{s\kappa\mu}{\sqrt{m}} \sqrt{\frac{\mu^2 s K \log^5 m}{m}} \text{ and} \tag{8.77}$$

$$\max_{1 \leq i \leq m} \left| \boldsymbol{b}_l^* \widetilde{\boldsymbol{h}}_i^0 \right| \cdot \left\| \boldsymbol{h}_i^\natural \right\|_2^{-1} \leq C_4 \frac{\mu \log^2 m}{\sqrt{m}}. \tag{8.78}$$

Proof With a similar strategy as in [13, Section C.6], we first show that the normalized singular vectors of $\boldsymbol{M}_i$ and $\boldsymbol{M}_i^{(l)}$, $i = 1, \ldots, s$ are close enough. We further extend this inequality to the scaled singular vectors, thereby converting the ℓ_2 metric to the distance function defined in (3.30). We finally prove the incoherence in terms of $\{\boldsymbol{b}_j\}_{j=1}^m$.

Recall that $\check{\boldsymbol{h}}_i^0$ and $\check{\boldsymbol{x}}_i^0$ are the leading left and right singular vectors of $\boldsymbol{M}_i$, $i = 1, \ldots, s$, and $\check{\boldsymbol{h}}_i^{0,(l)}$ and $\check{\boldsymbol{x}}_i^{0,(l)}$ are the leading left and right singular vectors of $\boldsymbol{M}_i^{(l)}$, $i = 1, \ldots, s$. By exploiting a variant of Wedin's sinΘ theorem [8, Therorem 2.1], we derive that

$$\leq \frac{c_1 \left\| \left(\boldsymbol{M}_i - \boldsymbol{M}_i^{(l)} \right) \check{\boldsymbol{x}}_i^{0,(l)} \right\|_2 + c_1 \left\| \check{\boldsymbol{h}}_i^{0,(l)*} \left(\boldsymbol{M}_i - \boldsymbol{M}_i^{(l)} \right) \right\|_2}{\sigma_1\left(\boldsymbol{M}_i^{(l)}\right) - \sigma_2(\boldsymbol{M}_i)}, \tag{8.79}$$

for $i = 1, \ldots, s$ with some constant $c_1 > 0$. According to [13, Section C.6], for $i = 1, \ldots, s$, we have

$$\sigma_1\left(\boldsymbol{M}_i^{(l)}\right) - \sigma_2(\boldsymbol{M}_i) \geq 3/4 - \left\| \boldsymbol{M}_i^{(l)} - \mathbb{E}[\boldsymbol{M}_i^{(l)}] \right\| - \|\boldsymbol{M}_i - \mathbb{E}[\boldsymbol{M}_i]\| \geq 1/2, \tag{8.80}$$

where the last step comes from [12, Lemma 6.16] provided that $m \gg (\mu^2 + \sigma^2)sK\log m$. As a result, we obtain that for $i = 1, \ldots, s$

$$\begin{aligned}&\left\|\beta_i^{0,(l)}\check{\boldsymbol{h}}_i^0 - \check{\boldsymbol{h}}_i^{0,(l)}\right\|_2 + \left\|\beta_i^{0,(l)}\check{\boldsymbol{x}}_i^0 - \check{\boldsymbol{x}}_i^{0,(l)}\right\|_2 \\ &\leq 2c_1\left\{\left\|\left(\boldsymbol{M}_i - \boldsymbol{M}_i^{(l)}\right)\check{\boldsymbol{x}}_i^{0,(l)}\right\|_2 + \left\|\check{\boldsymbol{h}}_i^{0,(l)*}\left(\boldsymbol{M}_i - \boldsymbol{M}_i^{(l)}\right)\right\|_2\right\},\end{aligned} \tag{8.81}$$

where

$$\beta_i^{0,(l)} := \underset{\alpha\in\mathbb{C},|\alpha|=1}{\operatorname{argmin}} \left\|\alpha\check{\boldsymbol{h}}_i^0 - \check{\boldsymbol{h}}_i^{0,(l)}\right\|_2 + \left\|\alpha\check{\boldsymbol{x}}_i^0 - \check{\boldsymbol{x}}_i^{0,(l)}\right\|_2. \tag{8.82}$$

It thus suffices to control the two terms on the right-hand side of (8.81). Therein,

$$\boldsymbol{M}_i - \boldsymbol{M}_i^{(l)} = \boldsymbol{b}_l\boldsymbol{b}_l^*\sum_{k=1}^{s}\boldsymbol{h}_k^\natural\boldsymbol{x}_k^{\natural*}\boldsymbol{a}_{kl}\boldsymbol{a}_{il}^* + e_l\boldsymbol{b}_l\boldsymbol{a}_{il}^*. \tag{8.83}$$

1. To bound the first term, we observe that

$$\begin{aligned}&\left\|\left(\boldsymbol{M}_i - \boldsymbol{M}_i^{(l)}\right)\check{\boldsymbol{x}}_i^{0,(l)}\right\|_2 \\ &= \left\|\boldsymbol{b}_l\boldsymbol{b}_l^*\sum_{k=1}^{s}\boldsymbol{h}_k^\natural\boldsymbol{x}_k^{\natural*}\boldsymbol{a}_{kl}\boldsymbol{a}_{il}^*\check{\boldsymbol{x}}_i^{0,(l)} + e_j\boldsymbol{b}_j\boldsymbol{a}_{il}^*\check{\boldsymbol{x}}_i^{0,(l)}\right\|_2 \\ &\leq \max_{1\leq k\leq s} s\|\boldsymbol{b}_l\|_2\cdot\left|\boldsymbol{b}_l^*\boldsymbol{h}_k^\natural\right|\cdot\left|\boldsymbol{a}_{kl}^*\boldsymbol{x}_i^\natural\right|\cdot\left|\boldsymbol{a}_{il}^*\check{\boldsymbol{x}}_i^{0,(l)}\right| + \|\boldsymbol{b}_l\|_2\cdot|e_l|\cdot\left|\boldsymbol{a}_{il}^*\check{\boldsymbol{x}}_i^{0,(l)}\right| \\ &\leq 30\frac{s\mu}{\sqrt{m}}\cdot\sqrt{\frac{K\log^2 m}{m}} + \frac{5\sigma^2}{m}\sqrt{\frac{K\log m}{m}},\end{aligned} \tag{8.84}$$

where we use the fact that $\|\boldsymbol{b}_l\|_2 = \sqrt{K/m}$, the incoherence condition, the bound (8.40), the assumption $|e_j| \leq \frac{\sigma^2}{m}$, and the condition that with probability exceeding $1 - O(m^{-10})$,

$$\max_{1\leq l\leq m}\left|\boldsymbol{a}_{il}^*\check{\boldsymbol{x}}_i^{0,(l)}\right| \leq 5\sqrt{\log m}, \tag{8.85}$$

due to the independence between $\check{\boldsymbol{x}}_i^{0,(l)}$ and $\boldsymbol{a}_{il}$ [13, Section C.6].

2. To control the second term, we observe that

$$\begin{aligned}
&\left\|\check{\boldsymbol{h}}_i^{0,(l)*}\left(\boldsymbol{M}_i-\boldsymbol{M}_i^{(l)}\right)\right\|_2 \\
&=\left\|\check{\boldsymbol{h}}_i^{0,(l)*}\boldsymbol{b}_l\boldsymbol{b}_l^*\sum_{k=1}^{s}\boldsymbol{h}_k^{\natural}\boldsymbol{x}_k^{\natural*}\boldsymbol{a}_{kl}\boldsymbol{a}_{il}^*+e_l\check{\boldsymbol{h}}_i^{0,(l)*}\boldsymbol{b}_l\boldsymbol{a}_{il}^*\right\|_2 \\
&\leq s\max_{1\leq k\leq s}\left\|\boldsymbol{a}_{il}^*\right\|_2\cdot\left|\boldsymbol{b}_l^*\boldsymbol{h}_k^{\natural}\right|\cdot\left|\boldsymbol{a}_{kl}^*\boldsymbol{x}_k^{\natural}\right|\cdot\left|\boldsymbol{b}_l^*\check{\boldsymbol{h}}_i^{0,(l)}\right|+\left\|\boldsymbol{a}_{il}^*\right\|_2\cdot|e_l|\cdot\left|\boldsymbol{b}_l^*\check{\boldsymbol{h}}_i^{0,(l)}\right| \\
&\leq\left(15\sqrt{\frac{\mu^2s^2K\log m}{m}}+3\sqrt{K}\frac{\sigma^2}{m}\right)\cdot\left(\left|\boldsymbol{b}_l^*\check{\boldsymbol{h}}_i^0\right|+\sqrt{\frac{K}{m}}\left\|\widetilde{\alpha}_i\check{\boldsymbol{h}}_i^0-\check{\boldsymbol{h}}_i^{0,(l)}\right\|_2\right),
\end{aligned} \tag{8.86}$$

where $|\widetilde{\alpha}_i|=1$. Here, the last step easily arises from the similar strategy used in (8.84) and [13, Section C.6]. Substitution of the bounds (8.84) and (8.86) into (8.81) yields

$$\begin{aligned}
&\left\|\beta_i^{0,(l)}\check{\boldsymbol{h}}_i^0-\check{\boldsymbol{h}}_i^{0,(l)}\right\|_2+\left\|\beta_i^{0,(l)}\check{\boldsymbol{x}}_i^0-\check{\boldsymbol{x}}_i^{0,(l)}\right\|_2 \\
&\leq 2C_1\left\{30\frac{\mu}{\sqrt{m}}\cdot\sqrt{\frac{s^2K\log^2 m}{m}}+\frac{5\sigma^2}{m}\sqrt{\frac{K\log m}{m}}\right. \\
&\quad\times\left(15\sqrt{\frac{\mu^2s^2K\log m}{m}}+3\sqrt{K}\frac{\sigma^2}{m}\right)\cdot \\
&\left.\left|\boldsymbol{b}_l^*\check{\boldsymbol{h}}_i^0\right|+\left(15\sqrt{\frac{\mu^2s^2K\log m}{m}}\sqrt{\frac{K}{m}}+3\sqrt{K}\frac{\sigma^2}{m}\right)\left\|\widetilde{\alpha}_i\check{\boldsymbol{h}}_i^0-\check{\boldsymbol{h}}_i^{0,(l)}\right\|_2\right\}.
\end{aligned} \tag{8.87}$$

Since the inequality (8.87) holds for any $|\widetilde{\alpha}_i|=1$, we can pick up $\widetilde{\alpha}_i=\beta^{0,(l)}$ and reformulate (8.87) as

$$\begin{aligned}
&\left(1-30c_1\sqrt{\frac{\mu^2s^2K\log m}{m}}\cdot\sqrt{\frac{K}{m}}-6\sqrt{K}\frac{\sigma^2}{m}\right)\cdot\left\|\beta_i^{0,(l)}\check{\boldsymbol{h}}_i^0-\check{\boldsymbol{h}}_i^{0,(l)}\right\|_2 \\
&\quad+\left\|\beta_i^{0,(l)}\check{\boldsymbol{x}}_i^0-\check{\boldsymbol{x}}_i^{0,(l)}\right\|_2 \\
&\quad\leq 60c_1\frac{\mu}{\sqrt{m}}\cdot\sqrt{\frac{s^2K\log^2 m}{m}}+\frac{10c_1\sigma^2}{m}\sqrt{\frac{K\log m}{m}} \\
&\qquad+\left(30c_1\sqrt{\frac{\mu^2s^2K\log m}{m}}+6c_1\sqrt{K}\frac{\sigma^2}{m}\right)\left|\boldsymbol{b}_l^*\check{\boldsymbol{h}}_i^0\right|.
\end{aligned} \tag{8.88}$$

With the assumption that $m \gg (\mu + \sigma^2)sK\log^{1/2} m$, it yields $1 - 30c_1\sqrt{\frac{\mu^2 s^2 K \log m}{m}} \cdot \sqrt{\frac{K}{m}} - 6\sqrt{K}\frac{\sigma^2}{m} \leq \frac{1}{2}$. Hence,

$$\begin{aligned}
&\max_{1\leq i\leq s, 1\leq j\leq m} \left\| \beta_i^{0,(l)} \check{\boldsymbol{h}}_i^0 - \check{\boldsymbol{h}}_i^{0,(l)} \right\|_2 + \left\| \beta_i^{0,(l)} \check{\boldsymbol{x}}_i^0 - \check{\boldsymbol{x}}_i^{0,(l)} \right\|_2 \\
&\leq 120c_1 \frac{\mu}{\sqrt{m}} \cdot \sqrt{\frac{s^2 K \log^2 m}{m}} + \frac{20c_1\sigma^2}{m}\sqrt{\frac{K\log m}{m}} \\
&\quad + \left(60c_1\sqrt{\frac{\mu^2 s^2 K \log m}{m}} + 12c_1\sqrt{K}\frac{\sigma^2}{m} \right) \cdot \max_{1\leq i\leq s, 1\leq j\leq m} \left| \boldsymbol{b}_l^* \check{\boldsymbol{h}}_i^0 \right|.
\end{aligned} \tag{8.89}$$

It thus suffices to control $\max_{1\leq i\leq s,1\leq j\leq m} |\boldsymbol{b}_l^* \check{\boldsymbol{h}}_i^0|$. Denote $\boldsymbol{M}_i \check{\boldsymbol{x}}^0 = \sigma_1(\boldsymbol{M}_i)\check{\boldsymbol{h}}_i^0$ and

$$\boldsymbol{W}_i = \sum_{j=1}^{m} \boldsymbol{b}_j \left(\sum_{k\neq i} \boldsymbol{b}_j^* \boldsymbol{h}_k^\natural \boldsymbol{x}_k^{\natural *} \boldsymbol{a}_{kj} + e_j \right) \boldsymbol{a}_{ij}^*, \tag{8.90}$$

which further leads to

$$\begin{aligned}
\left| \boldsymbol{b}_l^* \check{\boldsymbol{h}}_i^0 \right| &= \frac{1}{\sigma_1(\boldsymbol{M}_i)} \left| \boldsymbol{b}_l^* \boldsymbol{M}_i \check{\boldsymbol{x}}_i^0 \right| \\
&\overset{\text{(i)}}{\leq} 2 \left| \sum_{j=1}^{m} (\boldsymbol{b}_l^* \boldsymbol{b}_j)\, \boldsymbol{b}_j^* \boldsymbol{h}_i^\natural \boldsymbol{x}_i^{\natural *} \boldsymbol{a}_{ij} \boldsymbol{a}_{ij}^* \check{\boldsymbol{x}}_i^0 \right| + 2 \left| \boldsymbol{b}_l^* \boldsymbol{W}_i \check{\boldsymbol{x}}_i^0 \right| \\
&\overset{\text{(ii)}}{\leq} 2 \left(\sum_{j=1}^{m} |\boldsymbol{b}_l^* \boldsymbol{b}_j| \right) \max_{1\leq j\leq m} \left\{ \left| \boldsymbol{b}_j^* \boldsymbol{h}_i^\natural \right| \cdot \left| \boldsymbol{a}_{ij}^* \boldsymbol{x}_i^\natural \right| \cdot \left| \boldsymbol{a}_{ij}^* \check{\boldsymbol{x}}_i^0 \right| \right\} \\
&\quad + 2\|\boldsymbol{b}_l\|_2 \cdot \|\boldsymbol{W}_i\| \cdot \|\check{\boldsymbol{x}}_i^0\|_2 \\
&\overset{\text{(iii)}}{\leq} 2\sqrt{\frac{K}{m}} \cdot \|\boldsymbol{W}_i\| + 8\log m \cdot \frac{\mu}{\sqrt{m}} \cdot (5\sqrt{\log m}) \cdot \\
&\quad \max_{1\leq j\leq m} \left\{ \left| \boldsymbol{a}_j^* \check{\boldsymbol{x}}_i^{0,(j)} \right| + \|\boldsymbol{a}_{ij}\|_2 \left\| \beta_i^{0,(j)} \check{\boldsymbol{x}}_i^0 - \check{\boldsymbol{x}}_i^{0,(j)} \right\|_2 \right\} \\
&\overset{\text{(iv)}}{\leq} \sqrt{\frac{K}{m\log m}} + 200\frac{\mu\log^2 m}{\sqrt{m}} + 120\sqrt{\frac{\mu^2 K \log^3 m}{m}} \\
&\quad \cdot \max_{1\leq j\leq m} \left\| \beta_i^{0,(j)} \check{\boldsymbol{x}}_i^0 - \check{\boldsymbol{x}}_i^{0,(j)} \right\|_2,
\end{aligned} \tag{8.91}$$

where $\beta_i^{0,(j)}$ is defined in (8.82). Here, (i) arises from the low bound $\sigma_1(\boldsymbol{M}_i) \geq \frac{1}{2}$ and the triangle inequality. (ii) uses the Cauchy–Schwarz inequality. The step (iii) comes from combining the incoherence condition, the bound (8.40), the triangle inequality, the estimate: $\sum_{j=1}^{m} |\boldsymbol{b}_l^* \boldsymbol{b}_j| \leq 4\log m$ [13, Lemma 48], $\|\boldsymbol{b}_l\| = \sqrt{K/m}$ and $\|\check{\boldsymbol{x}}_i^0\|_2 = 1$. The last step (iv) exploits the inequality (8.85) to yield that with probability $1 - O(m^{-9})$ [12],

$$\|\boldsymbol{W}_i\| \leq \frac{1}{2\sqrt{\log m}}, \tag{8.92}$$

if $m \gg (\mu^2 + \sigma^2) sK \log^2 m$. The bound (8.91) further leads to

$$\max_{1\leq i\leq s} \left|\boldsymbol{b}_l^* \check{\boldsymbol{h}}_i^0\right| \leq \sqrt{\frac{K}{m\log m}} + 200\frac{\mu\log^2 m}{\sqrt{m}} + 120\sqrt{\frac{\mu^2 K \log^3 m}{m}} \cdot \max_{1\leq i\leq s, 1\leq j\leq m} \left\|\beta_i^{0,(j)} \check{\boldsymbol{x}}_i^0 - \check{\boldsymbol{x}}_i^{0,(j)}\right\|_2. \tag{8.93}$$

Combining the bound (8.89) and (8.93) gives

$$\begin{aligned}
&\max_{1\leq i\leq s, 1\leq l\leq m} \left\|\beta_i^{0,(l)} \check{\boldsymbol{h}}_i^0 - \check{\boldsymbol{h}}_i^{0,(l)}\right\|_2 + \left\|\beta_i^{0,(l)} \check{\boldsymbol{x}}_i^0 - \check{\boldsymbol{x}}_i^{0,(l)}\right\|_2 \\
&\leq 120 c_1 \frac{\mu}{\sqrt{m}} \cdot \sqrt{\frac{s^2 K \log^2 m}{m}} + \frac{20 c_1 \sigma^2}{m}\sqrt{\frac{K\log m}{m}} \\
&\quad + \left(60 c_1 \sqrt{\frac{\mu^2 s^2 K \log m}{m}} + 12 c_1 \sqrt{K}\frac{\sigma^2}{m}\right) \cdot \\
&\quad \left(\sqrt{\frac{K}{m\log m}} + 200\frac{\mu \log^2 m}{\sqrt{m}} + 120\sqrt{\frac{\mu^2 K\log^3 m}{m}} \right. \\
&\quad \left. \max_{1\leq i\leq s, 1\leq j\leq m} \left\|\beta_i^{0,(j)} \check{\boldsymbol{x}}_i^0 - \check{\boldsymbol{x}}_i^{0,(j)}\right\|_2\right).
\end{aligned} \tag{8.94}$$

As long as $m \gg (\mu^2 + \sigma^2) s^2 K \log^2 m$, we have

$$\left(60 c_1 \sqrt{\frac{\mu^2 s^2 K \log m}{m}} + 12 c_1 \sqrt{K}\frac{\sigma^2}{m}\right) \cdot 120\sqrt{\frac{\mu^2 s^2 K \log^3 m}{m}} \leq 1/2. \tag{8.95}$$

Reformulating the inequality (8.94), we have

$$\max_{1\leq i\leq s,1\leq l\leq m}\left\|\beta_i^{0,(l)}\check{\boldsymbol{h}}_i^0-\check{\boldsymbol{h}}_i^{0,(l)}\right\|_2+\left\|\beta_i^{0,(l)}\check{\boldsymbol{x}}_i^0-\check{\boldsymbol{x}}_i^{0,(l)}\right\|_2$$
$$\leq C_4\frac{\mu}{\sqrt{m}}\sqrt{\frac{\mu^2s^2K\log^5 m}{m}}, \tag{8.96}$$

for some constant $C_4>0$. Taking the bound (8.96) together with (8.93), it yields

$$\max_{1\leq i\leq s,1\leq l\leq m}\left|\boldsymbol{b}_l^*\check{\boldsymbol{h}}_i^0\right|\leq\sqrt{\frac{K}{m\log m}}+200\frac{\mu\log^2 m}{\sqrt{m}}+120\sqrt{\frac{\mu^2K\log^3 m}{m}}.$$
$$\max_{1\leq i\leq s,1\leq j\leq m}\quad C_4\frac{\mu}{\sqrt{m}}\sqrt{\frac{\mu^2s^2K\log^5 m}{m}}$$
$$\leq\ c_2\frac{\mu\log^2 m}{\sqrt{m}}, \tag{8.97}$$

for some constant $c_2>0$, as long as $m\gg(\mu^2+\sigma^2)sK\log^2 m$.

We further scale the preceding bounds to the final version. Based on [13, Section C.6], one has

$$\left\|\alpha\boldsymbol{h}^0-\boldsymbol{h}^{0,(l)}\right\|_2+\left\|\alpha\boldsymbol{x}^0-\boldsymbol{x}^{0,(l)}\right\|_2$$
$$\leq\left\|\left(\boldsymbol{M}_i-\boldsymbol{M}_i^{(l)}\right)\check{\boldsymbol{x}}_i^{0,(l)}\right\|_2+6\left\{\left\|\alpha\check{\boldsymbol{h}}_i^0-\check{\boldsymbol{h}}_i^{0,(l)}\right\|_2+\left\|\alpha\check{\boldsymbol{x}}_i^0-\check{\boldsymbol{x}}_i^{0,(l)}\right\|_2\right\}. \tag{8.98}$$

Taking the bounds (8.84), (8.96), and (8.98) collectively yields

$$\min_{\alpha_i\in\mathbb{C},|\alpha_i|=1}\left\|\alpha_i\boldsymbol{h}_i^0-\boldsymbol{h}_i^{0,(l)}\right\|_2+\left\|\alpha_i\boldsymbol{x}_i^0-\boldsymbol{x}_i^{0,(l)}\right\|_2\leq c_5\frac{\mu}{\sqrt{m}}\sqrt{\frac{\mu^2s^2K\log^5 m}{m}}, \tag{8.99}$$

for some constant $c_5>0$, as long as $m\gg(\mu^2+\sigma^2)s^2K\log^2 m$.

Furthermore, by exploiting the technical methods provided in [13, Section C.6], we have

$$\begin{aligned}
\operatorname{dist}\left(z^{0,(l)}, \widetilde{z}^0\right) &= \min_{\alpha_i \in \mathbb{C}} \sqrt{\sum_{i=1}^{s} \left\| \frac{1}{\overline{\alpha_i}} \boldsymbol{h}^{0,(l)} - \frac{1}{\overline{\alpha_i^0}} \boldsymbol{h}^0 \right\|_2^2 + \left\| \alpha_i \boldsymbol{x}^{0,(l)} - \alpha_i^0 \boldsymbol{x}^0 \right\|_2^2} \\
&\overset{\text{(i)}}{\leq} \min_{\alpha_i \in \mathbb{C}, |\alpha_i|=1} \sqrt{\sum_{i=1}^{s} \left\| \frac{1}{\overline{\alpha_i^0}} \boldsymbol{h}^0 - \frac{\alpha_i}{\overline{\alpha_i^0}} \boldsymbol{h}^{0,(l)} \right\|_2^2 + \left\| \alpha_i^0 \boldsymbol{x}^0 - \alpha_i \alpha_i^0 \boldsymbol{x}^{0,(l)} \right\|_2^2} \\
&\overset{\text{(ii)}}{\leq} 2\sqrt{s} \min_{\alpha_i \in \mathbb{C}, |\alpha_i|=1} \left\{ \left\| \boldsymbol{h}_i^0 - \alpha_i \boldsymbol{h}_i^{0,(l)} \right\|_2 + \left\| \boldsymbol{x}_i^0 - \alpha_i \boldsymbol{x}_i^{0,(l)} \right\|_2 \right\} \\
&\leq 2c_5 \frac{s\mu}{\sqrt{m}} \sqrt{\frac{\mu^2 s K \log^5 m}{m}},
\end{aligned} \tag{8.100}$$

where α_i^0 is defined in (3.30) and satisfies

$$\frac{1}{2} \leq |\alpha_i^0| \leq 2. \tag{8.101}$$

Here, the step (i) occurs since the feasible set for the latter optimization problem is smaller, and (ii) follows directly from [13, Lemma 19], [13, Lemma 52]. This accomplishes the proof for the claim (8.77). We further move to the proof for the claim (8.78).

In terms of $|\boldsymbol{b}_l^* \widetilde{\boldsymbol{h}}_i^0|$, one has

$$\left| \boldsymbol{b}_l^* \widetilde{\boldsymbol{h}}_i^0 \right| \leq \left| \boldsymbol{b}_l^* \frac{1}{\overline{\alpha_i^0}} \boldsymbol{h}_i^0 \right| \leq \left| \frac{1}{\overline{\alpha_i^0}} \right| \left| \boldsymbol{b}_l^* \boldsymbol{h}_i^0 \right| \leq 2 \left| \sqrt{\sigma_1(\boldsymbol{M}_i)} \boldsymbol{b}_l^* \check{\boldsymbol{h}}_i^0 \right| \leq 2\sqrt{2} c_2 \frac{\mu \log^2 m}{\sqrt{m}}, \tag{8.102}$$

based on fact that

$$\frac{1}{2} \leq \sigma_1(\boldsymbol{M}_i) \leq 2. \tag{8.103}$$

Lemma 8.8 *Suppose the sample complexity* $m \gg (\mu^2 + \sigma^2) s^{3/2} K \log^5 m$. *Then with probability at least* $1 - O(m^{-9})$,

$$\max_{1 \leq i \leq s, 1 \leq j \leq m} \left| \boldsymbol{a}_{ij}^* \left(\widetilde{\boldsymbol{x}}_i^0 - \boldsymbol{x}_i^\natural \right) \right| \cdot \left\| \boldsymbol{x}_i^\natural \right\|_2^{-1} \leq C_3 \frac{1}{\sqrt{s} \log^{3/2} m}. \tag{8.104}$$

Proof Recall several alignment parameters defined before:

$$\alpha_i^0 := \operatorname*{argmin}_{\alpha\in\mathbb{C}} \left\|\frac{1}{\overline{\alpha}}\boldsymbol{h}_i^0 - \boldsymbol{h}_i^\natural\right\|_2^2 + \left\|\alpha\boldsymbol{x}_i^0 - \boldsymbol{x}_i^\natural\right\|_2^2,$$

$$\alpha_i^{0,(l)} := \operatorname*{argmin}_{\alpha\in\mathbb{C}} \left\|\frac{1}{\overline{\alpha}}\boldsymbol{h}_i^{0,(l)} - \boldsymbol{h}_i^\natural\right\|_2^2 + \left\|\alpha\boldsymbol{x}_i^{0,(l)} - \boldsymbol{x}_i^\natural\right\|_2,$$

$$\alpha_{i,\text{mutual}}^{0,(l)} := \operatorname*{argmin}_{\alpha\in\mathbb{C}} \left\|\frac{1}{\overline{\alpha}}\boldsymbol{h}_i^{0,(l)} - \frac{1}{\overline{\alpha_i^0}}\boldsymbol{h}_i^0\right\|_2^2 + \left\|\alpha\boldsymbol{x}_i^{0,(l)} - \alpha_i^0\boldsymbol{x}_i^0\right\|_2^2.$$

Combining (8.17a) and (8.100) with the triangle inequality yields that

$$\begin{aligned}
&\leq \sqrt{\left\|\frac{1}{\overline{\alpha_{i,\text{mutual}}^{0,(l)}}}\boldsymbol{h}_i^{0,(l)} - \frac{1}{\overline{\alpha_i^0}}\boldsymbol{h}_i^0\right\|_2^2 + \left\|\alpha_{i,\text{mutual}}^{0,(l)}\boldsymbol{x}_i^{0,(l)} - \alpha_i^0\boldsymbol{x}_i^0\right\|_2^2}\\
&\leq 2c_5\frac{s\mu}{\sqrt{m}}\sqrt{\frac{\mu^2 K\log^5 m}{m}} + C_1\frac{1}{\sqrt{s}\log^2 m}\\
&\leq 2C_1\frac{1}{\sqrt{s}\log^2 m},
\end{aligned} \tag{8.105}$$

where the last step holds as long as $m \gg (\mu^2+\sigma^2)s\sqrt{sK}\log^{9/2} m$.

According to [13, Section C.7], [13, Lemma 55], and the bound (8.100), it implies that

$$\begin{aligned}
&\sqrt{\left\|\frac{1}{\overline{\alpha_i^{0,(l)}}}\boldsymbol{h}_i^{0,(l)} - \frac{1}{\overline{\alpha_i^0}}\boldsymbol{h}_i^0\right\|_2^2 + \left\|\alpha_i^{0,(l)}\boldsymbol{x}_i^{0,(l)} - \alpha_i^0\boldsymbol{x}_i^0\right\|_2^2}\\
&\lesssim \sqrt{\left\|\frac{1}{\overline{\alpha_i^0}}\boldsymbol{h}_i^0 - \frac{1}{\overline{\alpha_{i,\text{mutual}}^{0,(l)}}}\boldsymbol{h}_i^{0,(l)}\right\|_2^2 + \left\|\alpha_i^0\boldsymbol{x}_i^0 - \alpha_{i,\text{mutual}}^{0,(l)}\boldsymbol{x}_i^{0,(l)}\right\|_2^2}\\
&\lesssim \frac{s\mu}{\sqrt{m}}\sqrt{\frac{\mu^2 K\log^5 m}{m}}.
\end{aligned} \tag{8.106}$$

Based on the above estimate, we can show that with high probability,

$$\begin{aligned}
&\left|\boldsymbol{a}_{il}^*\left(\alpha_i^0\boldsymbol{x}_i^0 - \boldsymbol{x}_i^\natural\right)\right|\\
&\overset{\text{(i)}}{\leq} \left|\boldsymbol{a}_{il}^*\left(\alpha_i^{0,(l)}\boldsymbol{x}_i^{0,(l)} - \boldsymbol{x}_i^\natural\right)\right| + \left|\boldsymbol{a}_{il}^*\left(\alpha_i^0\boldsymbol{x}_i^0 - \alpha_i^{0,(l)}\boldsymbol{x}_i^{0,(l)}\right)\right|
\end{aligned}$$

$$\overset{\text{(ii)}}{\leq} 5\sqrt{\log m}\left\|\boldsymbol{a}_{il}^*\left(\alpha^{0,(l)}\boldsymbol{x}_i^{0,(l)} - \boldsymbol{x}_i^{\natural}\right)\right\|_2 + \|\boldsymbol{a}_{il}\|_2\left\|\boldsymbol{a}_{il}^*\left(\alpha_i^0\boldsymbol{x}_i^0 - \alpha_i^{0,(l)}\boldsymbol{x}_i^{0,(l)}\right)\right\|_2$$

$$\overset{\text{(iii)}}{\lesssim} \sqrt{\log m}\cdot\frac{1}{\sqrt{s}\log^2 m} + \sqrt{K}\frac{s\mu}{\sqrt{m}}\sqrt{\frac{\mu^2 K\log^5 m}{m}}$$

$$\overset{\text{(iv)}}{\lesssim} \frac{1}{\sqrt{s}\log^{3/2} m}, \tag{8.107}$$

where (i) arises from the triangle inequality, (ii) uses Cauchy–Schwarz inequality and the independence between $\boldsymbol{x}_i^{0,(l)}$ and $\boldsymbol{a}_{il}$, (iii) holds since (8.106), and (iv) occurs as long as $m \gg (\mu^2+\sigma^2)s^{3/2}K\log^4 m$.

8.4 Theoretical Analysis of Wirtinger Flow with Random Initialization for Blind Demixing

Based on the notations for blind demixing introduced in Sect. 3.4.2, we present the theoretical analysis of Wirtinger flow with random initialization in this section, which is based on [6]. It demonstrates that random initialization which enjoys a model-agnostic and natural initialization implementation for practitioners is good enough to guarantee Wirtinger flow to linearly converge to the optimal solution.

To present the theorem, we begin with introducing some notations. Let $\widetilde{\boldsymbol{h}}_i^t$ and $\widetilde{\boldsymbol{x}}_i^t$ be

$$\widetilde{\boldsymbol{h}}_i^t = \frac{1}{\overline{\omega_i^t}}\boldsymbol{h}_i^t \quad \text{and} \quad \widetilde{\boldsymbol{x}}_i^t = \omega_i^t\boldsymbol{x}_i^t, \tag{8.108}$$

for $i = 1,\ldots,s$, respectively, where alignment parameters are denoted as ω_i. Without loss of the generality, we assume that the ground truth $\boldsymbol{x}_i^{\natural} = q_i\boldsymbol{e}_1$ for $i = 1,\ldots,s$, where $0 < q_i \leq 1, i = 1,\ldots,s$ are some constants and define a parameter $\kappa = \frac{\max_i q_i}{\min_i q_i}$. For simplification, for $i = 1,\ldots,s$, the first entry and the rest entries of $\boldsymbol{x}_i^t$ are denoted as

$$x_{i1}^t \quad \text{and} \quad \boldsymbol{x}_{i\perp}^t := \left[x_{ij}^t\right]_{2\leq j\leq N}, \tag{8.109}$$

respectively. Hence, (8.113) and (8.114) can be reformulated as

$$\alpha_{x_i} := \widetilde{x}_{i1}^t \quad \text{and} \quad \beta_{x_i} := \left\|\widetilde{\boldsymbol{x}}_{i\perp}^t\right\|_2. \tag{8.110}$$

Define the norm of signal component and the perpendicular component in terms of $\boldsymbol{h}_i$ for $i = 1, \ldots, s$, as

$$\alpha_{\boldsymbol{h}_i} := \left\langle \boldsymbol{h}_i^\natural, \widetilde{\boldsymbol{h}}_i^t \right\rangle / \left\| \boldsymbol{h}_i^\natural \right\|_2, \tag{8.111}$$

$$\beta_{\boldsymbol{h}_i} := \left\| \widetilde{\boldsymbol{h}}_i^t - \frac{\left\langle \boldsymbol{h}_i^\natural, \widetilde{\boldsymbol{h}}_i^t \right\rangle}{\left\| \boldsymbol{h}_i^\natural \right\|_2^2} \boldsymbol{h}_i^\natural \right\|_2, \tag{8.112}$$

respectively. Likewise, the norms of the signal component and the perpendicular component in terms of $\boldsymbol{x}_i$ for $i = 1, \ldots, s$ are given by

$$\alpha_{\boldsymbol{x}_i} := \left\langle \boldsymbol{x}_i^\natural, \widetilde{\boldsymbol{x}}_i^t \right\rangle / \left\| \boldsymbol{x}_i^\natural \right\|_2, \tag{8.113}$$

$$\beta_{\boldsymbol{x}_i} := \left\| \widetilde{\boldsymbol{x}}_i^t - \frac{\left\langle \boldsymbol{x}_i^\natural, \widetilde{\boldsymbol{x}}_i^t \right\rangle}{\left\| \boldsymbol{x}_i^\natural \right\|_2^2} \boldsymbol{x}_i^\natural \right\|_2, \tag{8.114}$$

respectively.

Theorem 8.3 ([6]) *Assuming that the initial points are randomly generated as (3.33), and the stepsize $\eta > 0$ obeys $\eta \asymp s^{-1}$. Suppose that the sample size satisfies*

$$m \geq C\mu^2 s^2 \kappa^4 \max\{K, N\} \log^{12} m$$

for some sufficiently large constant $C > 0$. Then with probability at least $1 - c_1 m^{-\nu} - c_1 m e^{-c_2 N}$ with some constants $\nu, c_1, c_2 > 0$, for a sufficiently small constant $0 \leq \gamma \leq 1$ and $T_\gamma \lesssim s \log(\max\{K, N\})$, it holds that

1. The randomly initialized Wirtinger flow linearly converges to $\boldsymbol{z}^\natural$, i.e.,

$$\operatorname{dist}(\boldsymbol{z}^t, \boldsymbol{z}^\natural) \leq \gamma \left(1 - \frac{\eta}{16\kappa}\right)^{t - T_\gamma} \|\boldsymbol{z}^\natural\|2, \ t \geq T_\gamma,$$

2. The magnitude ratios of the signal component to the perpendicular component in terms of $\boldsymbol{h}_i^t$ and $\boldsymbol{x}_i^t$ obey

$$\max_{1 \leq i \leq s} \frac{\alpha_{\boldsymbol{h}_i^t}}{\beta_{\boldsymbol{h}_i^t}} \gtrsim \frac{1}{\sqrt{K \log K}} (1 + c_3 \eta)^t, \tag{8.115a}$$

$$\max_{1 \leq i \leq s} \frac{\alpha_{\boldsymbol{x}_i^t}}{\beta_{\boldsymbol{x}_i^t}} \gtrsim \frac{1}{\sqrt{N \log N}} (1 + c_4 \eta)^t, \tag{8.115b}$$

respectively, where $t = 0, 1, \cdots$ for some constant $c_3, c_4 > 0$.

The precise statistical analysis on the computational efficiency of Wirtinger flow with random initialization is illustrated in Theorem 8.3. In Stage I, it takes randomly initialized Wirtinger flow $T_\gamma = \mathscr{O}(s\log(\max\{K,N\}))$ iterations to reach a local region near the ground truth that enjoys strong convexity and strong smoothness. In Stage II, it takes $\mathscr{O}(s\log(1/\varepsilon))$ iterations to linearly converge ε-accurate point. Hence, the randomly initialized Wirtinger flow is guaranteed to converge to the ground truth with the iteration complexity $\mathscr{O}(s\log(\max\{K,N\})+s\log(1/\varepsilon))$ where the sample size is $m \gtrsim s^2\max\{K,N\}\text{poly}\log m$.

The proof of Theorem 8.3 is briefly summarized in the following. The key idea is to investigate the dynamics of the iterates of Wirtinger flow with random initialization.

1. **Stage I**:

- **Dynamics of population-level state evolution.** Establish the population-level state evolution of α_{x_i} (8.116a) and β_{x_i} (8.116b), α_{h_i} (8.117a), β_{h_i} (8.117b), respectively:

$$\alpha_{x_i^{t+1}} = (1-\eta)\alpha_{x_i^t} + \eta\frac{q_i\alpha_{h_i^t}}{\alpha_{h_i^t}^2+\beta_{h_i^t}^2}, \tag{8.116a}$$

$$\beta_{x_i^{t+1}} = (1-\eta)\beta_{x_i^t}. \tag{8.116b}$$

Similarly, the population-level state evolution for both $\alpha_{h_i^t}$ and $\beta_{h_i^t}$:

$$\alpha_{h_i^{t+1}} = (1-\eta)\alpha_{h_i^t} + \eta\frac{q_i\alpha_{x_i^t}}{\alpha_{x_i^t}^2+\beta_{x_i^t}^2}, \tag{8.117a}$$

$$\beta_{h_i^{t+1}} = (1-\eta)\beta_{h_i^t}, \tag{8.117b}$$

where the sample size approaches infinity. The approximate state evolution (8.118) is then established, which is significantly close to the population-level state evolution:

$$\alpha_{h_i^{t+1}} = \left(1-\eta+\frac{\eta q_i\gamma_{h_i^t}}{\alpha_{x_i^t}^2+\beta_{x_i^t}^2}\right)\alpha_{h_i^t} + \eta\left(1-_{h_i^t}\right)\frac{q_i\alpha_{x_i^t}}{\alpha_{x_i^t}^2+\beta_{x_i^t}^2}, \tag{8.118a}$$

$$\beta_{h_i^{t+1}} = \left(1-\eta+\frac{\eta q_i\varphi_{h_i^t}}{\alpha_{x_i^t}^2+\beta_{x_i^t}^2}\right)\beta_{h_i^t}, \tag{8.118b}$$

$$\alpha_{x_i^{t+1}} = \left(1 - \eta + \frac{\eta q_i \gamma_{x_i^t}}{\alpha_\upsilon \boldsymbol{h}_i^{t\,2} + \beta_{h_i^t}^2}\right) \alpha_{x_i^t} + \eta \left(1 - \upsilon_{x_i^t}\right) \frac{q_i \alpha_{h_i^t}}{\alpha_{h_i^t}^2 + \beta_{h_i^t}^2}, \tag{8.118c}$$

$$\beta_{x_i^{t+1}} = \left(1 - \eta + \frac{\eta q_i \varphi_{x_i^t}}{\alpha_{h_i^t}^2 + \beta_{h_i^t}^2}\right) \beta_{x_i^t}, \tag{8.118d}$$

where the perturbation terms are denoted as $\{\gamma_{h_i^t}\}$, $\{\gamma_{x_i^t}\}$, $\{\varphi_{h_i^t}\}$, $\{\varphi_{x_i^t}\}$, $\{\upsilon_{h_i^t}\}$, and $\{\upsilon_{x_i^t}\}$.

- **Dynamics of approximate state evolution.** Show that if α_{h_i} (8.111), β_{h_i} (8.112), α_{x_i} (8.113), and β_{x_i} (8.114) obey the approximate state evolution (8.118), it has some $T_\gamma = \mathcal{O}(s \log(\max\{K, N\}))$ such that $\mathrm{dist}(z^{T_\gamma}, z^\natural) \leq \gamma$. Furthermore, the ratio α_{h_i}/β_{h_i} and α_{x_i}/β_{x_i} enjoys exponential growth.
- **Leave-one-out arguments.** Identify the conditions where α_{h_i}, β_{h_i}, α_{x_i}, and β_{x_i} obey the approximate state evolution (8.118) with high probability, followed by demonstrating the iterates of randomly initialized Wirtinger flow that solve the blind demixing problem satisfy the conditions.

2. **Stage II: Local geometry in the region of incoherence and contraction.** Invoke the prior theory provided in [5] to show local convergence of the random initialized Wirtinger flow in Stage II.

8.5 The Basic Concepts on Riemannian Optimization

As a supplementary of Sect. 3.4.3, we introduce some basic concepts on Riemannian optimization via some examples. More details can be referred to the book [1].

Embedded Submanifolds Denote $\mathcal{N}$ as a subset of a manifold $\mathcal{M}$, which $\mathcal{N}$ admits at most one differentiable structure [1, Section 3.3]. Some examples of embedded submanifold are provided in the following.

Example 8.1 (The Stiefel Manifold) The Stiefel manifold is an embedded submanifold of $\mathbb{R}^{m\times n}$. For $n \leq m$, a Stiefel manifold can be denoted as

$$\mathrm{St}(n, m) := \left\{\boldsymbol{X} \in \mathbb{R}^{m\times n} : \boldsymbol{X}^\top \boldsymbol{X} = \boldsymbol{I}_n\right\}, \tag{8.119}$$

which is the set of all $m \times n$ orthonormal matrices, where $\boldsymbol{I}_n$ denotes the $n \times n$ identity matrix. For $n = 1$, the Stiefel manifold $\mathrm{St}(n, m)$ reduces to the unit sphere $\mathbb{S}^{m-1}$.

Tangent Vectors and Tangent Spaces Denote $\mathfrak{F}_{\boldsymbol{x}}(\mathcal{M})$ as the set of smooth real-valued functions. A tangent vector $\xi_{\boldsymbol{x}}$ to a manifold $\mathcal{M}$ at a point $\boldsymbol{x}$ is a mapping from $\mathfrak{F}_{\boldsymbol{x}}(\mathcal{M})$ to $\mathbb{R}$. For $f \in \mathfrak{F}_{\boldsymbol{x}}(\mathcal{M})$, there exists a curve ϕ on $\mathcal{M}$ with $\phi(0) = x$, such that

$$\xi_x f = \left.\frac{\mathrm{d}(f(\phi(t)))}{\mathrm{d}t}\right|_{t=0}. \tag{8.120}$$

Based on the curve ϕ, it yields a straightforward identification of the tangent space $T_{\boldsymbol{x}}\mathcal{M}$ with the set

$$\{\phi'(0) : \phi \text{ curve in } \mathcal{M}, \phi(0) = x\}. \tag{8.121}$$

An example of tangent space for sphere is presented in the following.

Example 8.2 (Tangent Space to a Sphere) Define a curve in the unit sphere S^{n-1} as $t \mapsto \gamma(t)$, and there is γ_0 at $t = 0$. Since $\gamma(t) \in S^{n-1}$ for all t, it holds that

$$\gamma^\top(t)\gamma(t) = 1 \tag{8.122}$$

for all t. Equation (8.122) is differentiated in terms of t, yielding

$$\dot{\gamma}^\top(t)\gamma(t) + \gamma^\top(t)\dot{\gamma}(t) = 0. \tag{8.123}$$

Thus $\dot{x}(0)$ is an entry of the set

$$\left\{\boldsymbol{x} \in \mathbb{R}^n : \gamma_0^\top \boldsymbol{x} = 0\right\}. \tag{8.124}$$

Furthermore, let $\boldsymbol{x}$ belong to the set (8.124). Then the curve

$$t \mapsto \gamma(t) := \frac{\gamma_0 + t\boldsymbol{x}}{\|\gamma_0 + t\boldsymbol{x}\|}$$

is on S^{n-1} and it holds $\dot{\gamma}(0) = \boldsymbol{x}$. Hence (8.124) is a subset of $T_{\gamma_0}S^{n-1}$, and

$$T_\gamma S^{n-1} = \left\{\boldsymbol{x} \in \mathbb{R}^n : \gamma^\top \boldsymbol{x} = 0\right\} \tag{8.125}$$

is the set of vectors orthogonal to the curve γ in $\mathbb{R}^n$.

Riemannian Metric As mentioned above, the notion of a directional derivative can be generalized by tangent vectors. To further identify which direction of act from $\boldsymbol{x}$ yields the steepest decrease in f, a notion of length with respect to tangent vectors is required. This can be achieved by assigning an inner product $\langle\cdot,\cdot\rangle_{\boldsymbol{x}}$, i.e., a symmetric positive-definite or bilinear operator, to each tangent space $T_{\boldsymbol{x}}\mathcal{M}$. The inner product $\langle\cdot,\cdot\rangle_{\boldsymbol{x}}$ for the point $\boldsymbol{x} \in \mathcal{M}$ is called the Riemannian metric, which can be represented as $g_{\boldsymbol{x}}$.

Product Manifolds The differentiable structure defined by two compact manifold $\mathcal{M}_1$ and $\mathcal{M}_2$, $\mathcal{M}_1 \times \mathcal{M}_2$ is called the product of the manifolds $\mathcal{M}_1$ and $\mathcal{M}_2$. Its manifold topology is equivalent to the product topology [1, Section 3.1.6], which means that the geometry concepts on the product manifolds can be represented by the set of elementwise geometry concepts on individual manifold. An example of product manifolds in the blind demixing problem (3.41) is introduced in the following.

Example 8.3 (Product Manifolds in the Blind Demixing Problem (3.41)) Taking the individual manifold $\mathcal{M}$ as an example, a smoothly varying inner product $g_{X}(\boldsymbol{\zeta}_{X}, \boldsymbol{\eta}_{X})$, where $\boldsymbol{\zeta}_{X}, \boldsymbol{\eta}_{X} \in T_{X}\mathcal{M}$, characterizes the notion of length that applies to each tangent space $T_{X}\mathcal{M}$. With a smoothly varying inner product g_{X}, the manifold $\mathcal{M}$ is called the *Riemannian manifold*, and the inner product is called the *Riemannian metric*. Denote $\mathcal{M}$ as the Riemannian manifold endowed with the Riemannian metric g_{X_k}, where $k \in [s]$ with $[s] = \{1, 2, \ldots, s\}$. The set of matrices $(\boldsymbol{X}_1, \ldots, \boldsymbol{X}_s)$ where $\boldsymbol{X}_k \in \mathcal{M}, k = 1, 2 \ldots, s$ is denoted as $\mathcal{M}^s = \underbrace{\mathcal{M} \times \mathcal{M} \times \cdots \times \mathcal{M}}_{s}$, and is called product manifold.

Based on the Riemannian geometry concepts, the notion of length on the product manifold can be characterized via endowing tangent space $T_{V}\mathcal{M}^s$ with the smoothly varying inner product, given by

$$g_{V}(\zeta_{V}, \eta_{V}) := \sum\nolimits_{k=1}^{s} g_{X_k}(\boldsymbol{\zeta}_{X_k}, \boldsymbol{\eta}_{X_k}), \tag{8.126}$$

where $\zeta_{V}, \eta_{V} \in T_{V}\mathcal{M}^s$ and $\boldsymbol{\zeta}_{X_k}, \boldsymbol{\eta}_{X_k} \in T_{X_k}\mathcal{M}$. Since $\mathcal{M}$ is the Riemannian manifold endowed with the Riemannian metric g_{X_k} for $\forall k \in [s]$, the product manifold $\mathcal{M}^s$ is also a Riemannian manifold, endowed with the Riemannian metric g_{V}.

Quotient Manifolds Computations related to subspaces are generally operated via representing the subspace by the span of corresponding matrices' columns. For a given subspace, to represent the subspace with a unique matrix, it is beneficial to divide the set of matrices into classes of "equivalent" elements that serve as the same object. This operation yields the geometry concept of quotient spaces, which is called *quotient manifolds* when concerning the Riemannian manifold optimization. We first present the general theory of quotient manifolds, then we introduce the corresponding representations of the blind demixing problem.

Denote a manifold endowed with an *equivalence relation* as $\sim\mathcal{M}$. Then the equivalence class containing $\boldsymbol{x}$ can be represented by the set

$$[\boldsymbol{x}] := \{\boldsymbol{y} \in \mathcal{M} : \boldsymbol{y} \sim \boldsymbol{x}\}, \tag{8.127}$$

which contains all elements that are equivalent to a point $\boldsymbol{x}$. The quotient of $\mathcal{M}$ by $\sim$ is defined as

$$\mathcal{M}/\sim := \{[\boldsymbol{x}] : \boldsymbol{x} \in \mathcal{M}\}, \tag{8.128}$$

which contains all equivalence classes of $\sim$ in $\mathcal{M}$. Thus, the points of $\mathcal{M}/\sim$ are subsets of $\mathcal{M}$, and the set $\mathcal{M}$ is called the total space of the quotient $\mathcal{M}/\sim$.

Furthermore, a natural projection that maps the elements in the manifold $\mathcal{M}$ to the quotient manifold $\mathcal{M}/\sim$ is defined as $\pi : \mathcal{M} \to \mathcal{M}/\sim$. If and only if $\boldsymbol{x} \sim \boldsymbol{y}$, it holds that $\pi(\boldsymbol{x}) = \pi(\boldsymbol{y})$ such that $[\boldsymbol{x}] = \pi^{-1}(\pi(\boldsymbol{x}))$.

An example of quotient manifold in the blind demixing problem (3.41) is provided in the following.

Example 8.4 (Quotient Manifold in the Blind Demixing Problem) According to the blind demixing problem given by (3.20):

$$\begin{aligned} &\text{find} \quad \text{rank}(\boldsymbol{W}_i) = 1, \text{ for } \boldsymbol{W}_1, \ldots, \boldsymbol{W}_s \\ &\text{subject to} \quad \left\| \sum\nolimits_{i=1}^{s} \mathcal{A}_i(\boldsymbol{W}_i) - \boldsymbol{y} \right\|_2 \le \varepsilon, \end{aligned}$$

this problem is a rank-constrained optimization problem. The key idea of Riemannian optimization for rank-constrained problem is based on matrix factorization [16, 23]. Specifically, the factorization $\boldsymbol{M}_k = \boldsymbol{w}_k \boldsymbol{w}_k^{\mathsf{H}}$ in problem (3.41) is established to identify rank-one Hermitian positive semidefinite matrices [22, 23]. Nevertheless, the factorization $\boldsymbol{M}_k = \boldsymbol{w}_k \boldsymbol{w}_k^{\mathsf{H}}$ is not unique since the transformation $\boldsymbol{w}_k \mapsto a_k \boldsymbol{w}_k$ with $a_k \in \{a_k \in \mathbb{C} : a_k a_k^* = a_k^* a_k = 1\}$ makes the matrix $\boldsymbol{w}_k \boldsymbol{w}_k^{\mathsf{H}}$ unchanged. To address this indeterminacy, the transformation $\boldsymbol{w}_k \mapsto a_k \boldsymbol{w}_k$ where $k = 1, 2, \ldots, s$, is embedded in an abstract search space, which constructs the equivalence class:

$$[\boldsymbol{M}_k] = \{a_k \boldsymbol{w}_k : a_k a_k^* = a_k^* a_k = 1, a_k \in \mathbb{C}\}. \tag{8.129}$$

The product of $[\boldsymbol{M}_k]$'s yields the equivalence class

$$[\boldsymbol{V}] = \{[\boldsymbol{M}_k]\}_{k=1}^{s}, \tag{8.130}$$

which is denoted as $\mathcal{M}^s/\sim$, called the *quotient space*. Since the quotient manifold $\mathcal{M}^s/\sim$ is an abstract space, the matrix representations defined in the computational space are needed to represent corresponding abstract geometric objects in the abstract space [1], thereby implementing the optimization algorithms. Denote an element of the quotient space $\mathcal{M}^s/\sim$ as $\tilde{\boldsymbol{V}}$ and its matrix representation in the computational space $\mathcal{M}^s$ as $\boldsymbol{V}$. Hence, there exists $\tilde{\boldsymbol{V}} = \pi(\boldsymbol{V})$ and $[\boldsymbol{V}] = \pi^{-1}(\pi(\boldsymbol{V}))$, where the mapping $\pi : \mathcal{M}^s \to \mathcal{M}^s/\sim$ is the natural projection.

Riemannian Submanifolds Denote an embedded submanifold of a Riemannian manifold $\overline{\mathcal{M}}$ as $\mathcal{M}$. Note that each tangent space $T_{\boldsymbol{x}}\mathcal{M}$ can be termed as a subspace of $T_{\boldsymbol{x}}\overline{\mathcal{M}}$. Hence, the Riemannian metric g on $\mathcal{M}$ can be derived by a Riemannian metric $\bar{g}$ of $\overline{\mathcal{M}}$ given by

$$g_{\boldsymbol{x}}(\boldsymbol{\eta}, \boldsymbol{\zeta}) = \bar{g}_{\boldsymbol{x}}(\boldsymbol{\eta}, \boldsymbol{\zeta}), \quad \boldsymbol{\eta}, \boldsymbol{\zeta} \in T_{\boldsymbol{x}}\mathcal{M}. \tag{8.131}$$

This implies that $\mathcal{M}$ is a Riemannian manifold. The manifold $\mathcal{M}$ with the Riemannian metric $g_{\boldsymbol{x}}$ is called a Riemannian submanifold of $\overline{\mathcal{M}}$. The orthogonal complement of the tangent space $T_{\boldsymbol{x}}\mathcal{M}$ in $T_{\boldsymbol{x}}\overline{\mathcal{M}}$ is called the normal space to $\mathcal{M}$ at $\boldsymbol{x}$, which is denoted by $(T_{\boldsymbol{x}}\mathcal{M})^{\perp}$:

$$(T_{\boldsymbol{x}}\mathcal{M})^{\perp} = \left\{\boldsymbol{\eta} \in T_{\boldsymbol{x}}\overline{\mathcal{M}} : g_{\boldsymbol{x}}(\boldsymbol{\eta}, \boldsymbol{\zeta}) = 0 \text{ for all } \boldsymbol{\zeta} \in T_{\boldsymbol{x}}\mathcal{M}\right\}. \tag{8.132}$$

Thus, the sum of an element of $T_{\boldsymbol{x}}\mathcal{M}$ and an element of $(T_{\boldsymbol{x}}\mathcal{M})^{\perp}$ can yield an element $\boldsymbol{\eta} \in T_{\boldsymbol{x}}\overline{\mathcal{M}}$:

$$\boldsymbol{\eta} = P_{\boldsymbol{x}}\boldsymbol{\eta} + P_{\boldsymbol{x}}^{\perp}\boldsymbol{\eta}, \tag{8.133}$$

where $P_{\boldsymbol{x}}^{\perp}$ is the orthogonal projection onto $(T_{\boldsymbol{x}}\mathcal{M})^{\perp}$ and $P_{\boldsymbol{x}}$ is the orthogonal projection onto $T_{\boldsymbol{x}}\mathcal{M}$. We first present a simple example, i.e., sphere S^{n-1} which is a Riemannian submanifold of $\mathbb{R}^n$. Based on the Riemannian submanifold of product manifolds, we further introduce the decomposition of the tangent space $T_{\boldsymbol{V}}\mathcal{M}^s$ of the product manifold $\mathcal{M}^s$ in the blind demixing problem.

Example 8.5 (Sphere) On the unit sphere S^{n-1} which is a Riemannian submanifold of $\mathbb{R}^n$, the inner product derived from the Euclidean inner product on $\mathbb{R}^n$ is represented as

$$\langle \boldsymbol{\xi}, \boldsymbol{\zeta}\rangle_{x} := \boldsymbol{\xi}^{\top}\boldsymbol{\zeta}. \tag{8.134}$$

The normal space is

$$\left(T_{\boldsymbol{x}}S^{n-1}\right)^{\perp} = \{\boldsymbol{x}\theta : \theta \in \mathbb{R}\}, \tag{8.135}$$

and the projections are given by $\mathrm{P}_{\boldsymbol{x}}\xi = \left(I - xx^{\top}\right)\xi, \quad \mathrm{P}_{\boldsymbol{x}}^{\perp}\xi = xx^{\top}\xi$ for $x \in S^{n-1}$.

Example 8.6 (Product Manifolds in Problem (3.41)*)* Considering the product manifold $\mathcal{M}^s$ endowed with the Riemannian metric (8.126), the tangent space $T_{\boldsymbol{V}}\mathcal{M}^s$ can be decomposed into two complementary vector spaces, given by Absil et al. [1]

$$T_{\boldsymbol{V}}\mathcal{M}^s = \mathcal{V}_{\boldsymbol{V}}\mathcal{M}^s \oplus \mathcal{H}_{\boldsymbol{V}}\mathcal{M}^s, \tag{8.136}$$

where $\oplus$ is the direct sum operator. Particularly, the set of vectors which are tangent to the set of equivalence class (8.130) is denoted as the *vertical space*, i.e., $\mathcal{V}_{\boldsymbol{V}}\mathcal{M}^s$. While the set of vectors which are orthogonal to the equivalence class (8.130) is denoted as the *horizontal space*, i.e., $\mathcal{H}_{\boldsymbol{V}}\mathcal{M}^s$. Hence the tangent space $T_{\tilde{\boldsymbol{V}}}(\mathcal{M}^s/\sim)$ at the point $\tilde{\boldsymbol{V}} \in \mathcal{M}^s/\sim$ can be represented by the horizontal space $\mathcal{H}_{\boldsymbol{V}}\mathcal{M}^s$ at point $\boldsymbol{V} \in \mathcal{M}^s$. Hence, the matrix representation of $\eta_{\tilde{\boldsymbol{V}}} \in T_{\tilde{\boldsymbol{V}}}(\mathcal{M}^s/\sim)$ [1, Section 3.5.8] can be represented by a unique element $\boldsymbol{\eta}_{\boldsymbol{V}} \in \mathcal{H}_{\boldsymbol{V}}\mathcal{M}^s$. Additionally, for each

$\xi_V, \eta_V \in T_V \mathscr{M}^s$, by exploiting the Riemannian metric $g_V(\zeta_V, \eta_V)$ (8.126),

$$g_{\tilde{V}}(\zeta_{\tilde{V}}, \eta_{\tilde{V}}) := g_V(\zeta_V, \eta_V) \tag{8.137}$$

defines a Rimannian metric on the quotient space $\mathscr{M}^s/\sim$ [1, Section 3.6.2], where $\zeta_{\tilde{V}}, \eta_{\tilde{V}} \in T_{\tilde{V}} \mathscr{M}^s$. With the Riemannian metric (8.137), the natural projection $\pi : \mathscr{M}^s \to \mathscr{M}^s/\sim$ is a mapping from the quotient manifold $\mathscr{M}^s/\sim$ to the computational space, which is also called *Riemannian submersion* $\mathscr{M}^s$ [1, Section 3.6.2]. According to the Riemannian submersion theory, the objects on the quotient manifold can be represented by corresponding objects in the computational space, which facilitates to develop Riemannian optimization algorithm on the Riemannian manifold.

8.6 Proof of Theorem 3.4

Based on the notions mentioned in Sect. 3.4.3.2, the Euclidean gradient of $f(\boldsymbol{v})$ in problem (3.41) with respect to $\boldsymbol{w}_k$ is given by

$$\nabla_{\boldsymbol{w}_k} f(\boldsymbol{v}) = 2 \cdot \sum_{i=1}^{L} \left(c_i \boldsymbol{J}_{ki} + c_i^* \boldsymbol{J}_{ki}^{\mathsf{H}} \right) \cdot \boldsymbol{w}_k, \tag{8.138}$$

where $c_i = \sum_{k=1}^{s} [\mathscr{J}_k(\boldsymbol{w}_k \boldsymbol{w}_k^{\mathsf{H}})]_i - y_i$. According to the definition of the horizontal space given in Table 3.4, it yields that $\nabla_{\boldsymbol{w}_k} f(\boldsymbol{v})$ is in the horizontal space due to $\nabla_{\boldsymbol{w}_k} f(\boldsymbol{v})^{\mathsf{H}} \boldsymbol{w}_k = \boldsymbol{w}_k^{\mathsf{H}} \nabla_{\boldsymbol{w}_k} f(\boldsymbol{v})$. Thus, the update rule in the Riemannian gradient descent algorithm, i.e., Algorithm 3.3, can be reformulated as

$$\boldsymbol{w}_k^{[t+1]} = \boldsymbol{w}_k^{[t]} - \frac{\alpha_t}{2 \left\| \boldsymbol{w}_k^{[t]} \right\|_2^2} \nabla_{\boldsymbol{w}_k} f(\boldsymbol{v}) |_{\boldsymbol{w}_k^{[t]}}, \tag{8.139}$$

according to the definition of the Riemannian metric $g_{\boldsymbol{w}_k}$ and the retraction $\mathscr{R}_{\boldsymbol{w}_k}$ in Table 3.4. The update rule can be reformulated as

$$\begin{bmatrix} \boldsymbol{w}_k^{[t+1]} \\ \overline{\boldsymbol{w}_k^{[t+1]}} \end{bmatrix} = \begin{bmatrix} \boldsymbol{w}_k^{[t]} \\ \overline{\boldsymbol{w}_k^{[t]}} \end{bmatrix} - \frac{\alpha_t}{\left\| \boldsymbol{w}_k^{[t]} \right\|_2^2} \begin{bmatrix} \frac{\partial f}{\partial \boldsymbol{w}_k^{\mathsf{H}}} |_{\boldsymbol{w}_k^{[t]}} \\ \overline{\frac{\partial f}{\partial \boldsymbol{w}_k^{\mathsf{H}}} |_{\boldsymbol{w}_k^{[t]}}} \end{bmatrix}, \tag{8.140}$$

according to the fact that $\nabla_{\boldsymbol{w}_k} f(\boldsymbol{v}) = 2 \frac{\partial f(\boldsymbol{v})}{\partial \boldsymbol{w}_k^{\mathsf{H}}}$.

The proof of Theorem 3.4 is summarized as follows.

- Lemma 8.9 characterizes the local geometry in the region of incoherence and contraction (RIC) illustrated in Definition 8.3, where the objective function $f(\boldsymbol{v})$ (3.41) enjoys restricted strong convexity and smoothness near the ground truth $\boldsymbol{v}^\natural$.

- Based on the property of the local geometry, Lemma 8.10 establishes the error contraction, i.e., convergence analysis.
- Lemma 8.11 demonstrates that the iterates of Algorithm 3.3, including the spectral initialization point, stay within the RIC. This is achieved by exploiting the induction arguments.

Definition 8.3 ($(\phi, \beta, \gamma, \boldsymbol{v}^{\natural}) - \mathscr{R}$ the Region of Incoherence and Contraction) Define $\boldsymbol{v}_i = [\boldsymbol{x}_i^{\mathsf{H}}\ \boldsymbol{h}_i^{\mathsf{H}}]^{\mathsf{H}} \in \mathbb{C}^{N+K}$ and $\boldsymbol{v} = [\boldsymbol{v}_1^{\mathsf{H}} \cdots \boldsymbol{v}_s^{\mathsf{H}}]^{\mathsf{H}} \in \mathbb{C}^{s(N+K)}$. For $\boldsymbol{v} \in (\phi, \theta, \gamma, \boldsymbol{v}^{\natural}) - \mathscr{R}$, there exists

$$\operatorname{dist}\left(\boldsymbol{v}^t, \boldsymbol{v}^{\natural}\right) \leq \phi, \tag{8.141a}$$

$$\max_{1\leq i\leq s, 1\leq j\leq m} \left|\boldsymbol{c}_{ij}^{\mathsf{H}}\left(\widetilde{\boldsymbol{x}}_i^t - \boldsymbol{x}_i^{\natural}\right)\right| \cdot \left\|\boldsymbol{x}_i^{\natural}\right\|_2^{-1} \leq C_2\theta, \tag{8.141b}$$

$$\max_{1\leq i\leq s, 1\leq j\leq m} \left|\boldsymbol{b}_j^{\mathsf{H}}\widetilde{\boldsymbol{h}}_i^t\right| \cdot \left\|\boldsymbol{h}_i^{\natural}\right\|_2^{-1} \leq C_3\gamma, \tag{8.141c}$$

where some constants $C_2, C_3 > 0$ and some sufficiently small constants $\phi, \theta, \gamma > 0$. Additionally, $\widetilde{\boldsymbol{h}}_i^t$ and $\widetilde{\boldsymbol{x}}_i^t$ are defined as $\widetilde{\boldsymbol{h}}_i^t = \frac{1}{\psi_i^t}\boldsymbol{h}_i^t$ and $\widetilde{\boldsymbol{x}}_i^t = \psi_i^t \boldsymbol{x}_i^t$ for $i = 1, \ldots, s$, with the alignment parameter ψ_i^t.

The Riemannian Hessian is denoted as $\operatorname{Hess} f(\boldsymbol{v}) := \operatorname{diag}(\{\operatorname{Hess}_{\boldsymbol{w}_i} f\}_{i=1}^s)$.

Lemma 8.9 ([7]) *Assuming a sufficiently small constant $\delta > 0$. If the number of measurements satisfies $m \gg \mu^2 s^2 \kappa^2 \max\{N, K\} \log^5 m$, then with probability exceeding $1 - \mathscr{O}(m^{-10})$, $\operatorname{Hess} f(\boldsymbol{v})$ satisfies*

$$\begin{aligned} & \boldsymbol{z}^{\mathsf{H}}\left[\boldsymbol{D}\operatorname{Hess} f(\boldsymbol{v}) + \operatorname{Hess} f(\boldsymbol{v})\boldsymbol{D}\right]\boldsymbol{z} \geq \frac{1}{4\kappa}\|\boldsymbol{z}\|_2^2 \\ & \textit{and} \quad \|\operatorname{Hess} f(\boldsymbol{v})\| \leq 2 + s \end{aligned} \tag{8.142}$$

simultaneously for all

$$\boldsymbol{z} = \left[\boldsymbol{z}_1^{\mathsf{H}} \cdots \boldsymbol{z}_s^{\mathsf{H}}\right]^{\mathsf{H}} \textit{ with } \boldsymbol{z}_i = \left[\left(\boldsymbol{x}_i - \boldsymbol{x}_i'\right)^{\mathsf{H}}\ \left(\boldsymbol{h}_i - \boldsymbol{h}_i'\right)^{\mathsf{H}}\ \left(\boldsymbol{x}_i - \boldsymbol{x}_i'\right)^{\top}\ \left(\boldsymbol{h}_i - \boldsymbol{h}_i'\right)^{\top}\right]^{\mathsf{H}},$$

and $\boldsymbol{D} = \operatorname{diag}\left(\{\boldsymbol{W}_i\}_{i=1}^s\right)$ with

$$\boldsymbol{W}_i = \operatorname{diag}\left(\left[\overline{\beta}_{i1}\boldsymbol{I}_K\ \overline{\beta}_{i2}\boldsymbol{I}_N\ \overline{\beta}_{i1}\boldsymbol{I}_K\ \overline{\beta}_{i2}\boldsymbol{I}_N\right]^*\right).$$

Here $\boldsymbol{v}$ is in the region $(\delta, \frac{1}{\sqrt{s}\log^{3/2} m}, \frac{\mu}{\sqrt{m}}\log^2 m, \boldsymbol{v}^{\natural}) - \mathscr{R}$, and one has

$$\max\{\left\|\boldsymbol{h}_i - \boldsymbol{h}_i^{\natural}\right\|_2, \left\|\boldsymbol{h}_i' - \boldsymbol{h}_i^{\natural}\right\|_2, \left\|\boldsymbol{x}_i - \boldsymbol{x}_i^{\natural}\right\|_2, \left\|\boldsymbol{x}_i' - \boldsymbol{x}_i^{\natural}\right\|_2\} \leq \delta/(\kappa\sqrt{s}),$$

for $i = 1, \ldots, s$ *and* $\mathbf{W}_i$*'s satisfy that for* $\beta_{i1}, \beta_{i2} \in \mathbb{R}$, *for* $i = 1, \ldots, s$

$$\max_{1\leq i\leq s} \max \left\{ \left|\beta_{i1} - \frac{1}{\kappa}\right|, \left|\beta_{i2} - \frac{1}{\kappa}\right| \right\} \leq \frac{\delta}{\kappa\sqrt{s}}.$$

Therein, $C_2, C_3 \geq 0$ *are numerical constants.*

Lemma 8.10 ([7]) *Assuming that the step size satisfies* $\alpha_t > 0$ *and* $\alpha_t \equiv \alpha \asymp s^{-1}$, *then with probability exceeding* $1 - \mathcal{O}(m^{-10})$,

$$\operatorname{dist}\left(\boldsymbol{v}^{t+1}, \boldsymbol{v}^{\natural}\right) \leq \left(1 - \frac{\alpha}{16\kappa}\right) \operatorname{dist}\left(\boldsymbol{v}^{t}, \boldsymbol{v}^{\natural}\right), \tag{8.143}$$

provided that the number of measurements follows $m \gg \mu^2 s^2 \kappa^4 \max\{N, K\} \log^5 m$ *and* $\boldsymbol{v}$ *is in the region* $(\delta, \frac{1}{\sqrt{s}\log^{3/2} m}, \frac{\mu}{\sqrt{m}} \log^2 m, \boldsymbol{v}^{\natural}) - \mathcal{R}$, *which is denoted as* $\mathcal{R}_{bd}$.

Lemma 8.11 ([7]) *Assuming the number of measurements*

$$m \gg \mu^2 s^2 \kappa^2 \max\{K, N\} \log^6 m,$$

then the spectral initialization point $\boldsymbol{v}^0$ *is in the region* $\mathcal{R}_{bd}$ *with probability exceeding* $1 - \mathcal{O}(m^{-9})$.

Assuming that t*-th iteration* $\boldsymbol{v}^t$ *is in the region* $\mathcal{R}_{bd}$ *and the number of measurements satisfy*

$$m \gg \mu^2 s^2 \kappa^2 \max\{K, N\} \log^8 m,$$

then with probability exceeding $1 - \mathcal{O}(m^{-9})$, *the* $(t+1)$*-th iteration* $\boldsymbol{v}^{t+1}$ *is also in the region* $\mathcal{R}_{bd}$, *which the step size satisfies* $\alpha_t > 0$ *and* $\alpha_t \equiv \alpha \asymp s^{-1}$.

8.7 Basic Concepts in Algebraic–Geometric Theory

In this section, we will introduce some basic concepts in algebraic–geometry theory, which contribute to the proof of Theorems 5.5 and 5.6. The content in this section is based on the paper [20].

8.7.1 Geometric Characterization of Dimension

Denote polynomials in $\mathbb{R}[\boldsymbol{x}] = \mathbb{R}[x_1, \ldots, x_n]$ as $f_1, \ldots, f_s$, their common root position $\mathcal{V}_{\mathbb{R}^n}(f_1, \ldots, f_s)$, called an *algebraic variety*, is defined as

$$\mathcal{V}_{\mathbb{R}^n}(f_1, \ldots, f_s) := \left\{\boldsymbol{\zeta} \in \mathbb{R}^n : f_i(\boldsymbol{\zeta}) = 0, \forall i \in [s]\right\}. \tag{8.144}$$

Considering the dimension of $\mathscr{V}_{\mathbb{R}^n}(f_1, \ldots, f_s)$, for the case of a single equation, i.e., $s = 1$, $\mathscr{V}_{\mathbb{R}^n}(f_1)$ is a hypersurface of $\mathbb{R}^n$ endowed with dimension $n - 1$; this is similar to the situation where a single linear equation identifies a linear subspace of dimension one less than the ambient dimension. There are other more complicated cases where $\mathscr{V}_{\mathbb{R}^n}(f_1)$ consists of a single point or no points, which is similar to algebra where a linear subspace has zero dimension only if it contains a single point or the origin point. In these cases, the common root position of the polynomials in the *algebraic closure* $\mathbb{C}$ of $\mathbb{R}$ is considered:

$$\mathscr{V}_{\mathbb{C}^n}(f_1, \ldots, f_s) := \left\{\boldsymbol{\zeta} \in \mathbb{C}^n : f_i(\boldsymbol{\zeta}) = 0, \forall i \in [s]\right\}. \tag{8.145}$$

The dimension of (8.145) $\mathscr{V}_{\mathbb{C}^n}(f_1, \ldots, f_s) \subset \mathbb{C}^n$ can be characterized by a well-developed theory [9, 11, 14]. The geometric characterization of $\dim \mathscr{V}_{\mathbb{C}^n}(f_1, \ldots, f_s)$ is presented by the following definition.

Definition 8.4 If $\mathscr{Y} = \mathscr{V}(g_1, \ldots, g_r)$ for some polynomials $g_1, \ldots, g_r \in \mathbb{C}[\boldsymbol{x}]$, $\mathscr{Y} \subset \mathbb{C}^n$ is defined to be closed. If $\mathscr{Y} = \mathscr{V}(g_1, \ldots, g_r)$ is not the union of two proper closed subsets, $\mathscr{Y}$ is defined to be irreducible. The dimension of geometric object $\mathscr{V}_{\mathbb{C}^n}(f_1, \ldots, f_s)$ is defined to be the largest non-negative integer d such that there is a chain of the form

$$\mathscr{V}_{\mathbb{C}^n}(f_1, \ldots, f_s) \supset \mathscr{Y}_0 \supsetneq \mathscr{Y}_1 \supsetneq \mathscr{Y}_2 \supsetneq \cdots \supsetneq \mathscr{Y}_d, \tag{8.146}$$

where $\mathscr{Y}_i$ for any $i \in \{1, \ldots, d\}$ is a closed irreducible subset of $\mathscr{V}_{\mathbb{C}^n}(f_1, \ldots, f_s)$.

Definition 8.4 can be termed as a generalization of the notion of dimension in linear algebra. For instance, considering that $\mathscr{Y}$ is a linear subspace of $\mathbb{C}^n$, $\dim \mathscr{Y}$ is the same as the maximal length of a descending chain of linear subspaces. The descending chain can be derived by removing a single basis vector of $\mathscr{Y}$ at each step. Please refer to Example 8.7 for details.

Example 8.7 Define a unit vector $\boldsymbol{e}_i$ with the value of 1 at position i zeros and zeros at the rest positions. Then $\mathscr{Y}_i = \text{Span}(\boldsymbol{e}_1, \ldots, \boldsymbol{e}_{n-i})$, $\mathbb{C}^n$ admits a chain

$$\mathbb{C}^n = \mathscr{Y}_0 \supsetneq \mathscr{Y}_1 \supsetneq \mathscr{Y}_2 \supsetneq \cdots \supsetneq \mathscr{Y}_{n-1} \supsetneq \mathscr{Y}_n := \{0\}. \tag{8.147}$$

Furthermore, the following propositions present several structural property about algebraic varieties.

Proposition 8.2 *Define $\mathscr{Y} = \mathscr{V}_{\mathbb{C}^n}(f_1, \ldots, f_s)$ for some $f_i \in \mathbb{C}[\boldsymbol{x}]$. With irreducible closed sets of $\mathbb{C}^n$ defined in Definition 8.4, i.e., $\mathscr{Y}_i$, $\mathscr{Y}$ can be represented as $\mathscr{Y} = \mathscr{Y}_1 \cup \cdots \cup \mathscr{Y}_d$ for some positive integer d. The set $\mathscr{Y}$ is minimal, that is, removing one of the $\mathscr{Y}_i$ would yield a union that is a strictly smaller set than $\mathscr{Y}$. The $\mathscr{Y}_i$ for $i \in \{1, \ldots, d\}$ are called the irreducible components of $\mathscr{Y}$.*

Definition 8.4 along with Proposition 8.2 demonstrate that the dimension of $\mathscr{V}_{\mathbb{C}^n}(f_1, \ldots, f_s)$ is zero if and only if the algebraic varieties $\mathscr{V}_{\mathbb{C}^n}(f_1, \ldots, f_s)$ consist of a finite number of points. These varieties are concerned in the paper [20].

Proposition 8.3 *Define $\mathscr{Y} = \mathscr{V}_{\mathbb{C}^n}(f_1, \ldots, f_s)$. Then the dimension of $\mathscr{Y}$ is 0 if and only if $\mathscr{Y}$ consists of a finite number of points of $\mathbb{C}^n$.*

References

1. Absil, P.A., Mahony, R., Sepulchre, R.: Optimization Algorithms on Matrix Manifolds. Princeton University Press, Princeton (2009)
2. Amelunxen, D., Bürgisser, P.: Intrinsic volumes of symmetric cones and applications in convex programming. Math. Program. **149**(1–2), 105–130 (2015)
3. Amelunxen, D., Lotz, M., McCoy, M.B., Tropp, J.A.: Living on the edge: phase transitions in convex programs with random data. Inf. Inference **3**(3), 224–294 (2014)
4. Chandrasekaran, V., Recht, B., Parrilo, P.A., Willsky, A.S.: The convex geometry of linear inverse problems. Found. Comput. Math. **12**(6), 805–849 (2012)
5. Dong, J., Shi, Y.: Nonconvex demixing from bilinear measurements. IEEE Trans. Signal Process. **66**(19), 5152–5166 (2018)
6. Dong, J., Shi, Y.: Blind demixing via Wirtinger flow with random initialization. In: The 22nd International Conference on Artificial Intelligence and Statistics (AISTATS), vol. 89, pp. 362–370 (2019)
7. Dong, J., Yang, K., Shi, Y.: Blind demixing for low-latency communication. IEEE Trans. Wireless Commun. **18**(2), 897–911 (2019)
8. Dopico, F.M.: A note on sin θ theorems for singular subspace variations. BIT Numer. Math. **40**(2), 395–403 (2000)
9. Eisenbud, D.: Commutative Algebra: With a View Toward Algebraic Geometry, vol. 150. Springer, Berlin (2013)
10. Fan, J., Wang, D., Wang, K., Zhu, Z., et al.: Distributed estimation of principal eigenspaces. Ann. Stat. **47**(6), 3009–3031 (2019)
11. Hartshorne, R.: Algebraic Geometry, vol. 52. Springer, Berlin (2013)
12. Ling, S., Strohmer, T.: Regularized gradient descent: a nonconvex recipe for fast joint blind deconvolution and demixing. Inf. Inference J. IMA **8**(1), 1–49 (2018)
13. Ma, C., Wang, K., Chi, Y., Chen, Y.: Implicit regularization in nonconvex statistical estimation: gradient descent converges linearly for phase retrieval, matrix completion and blind deconvolution. arXiv preprint:1711.10467 (2017)
14. Matsumura, H.: Commutative Ring Theory, vol. 8. Cambridge University, Cambridge (1989)
15. McCoy, M., Tropp, J.: Sharp recovery bounds for convex demixing, with applications. Found. Comput. Math. **14**(3), 503–567 (2014)
16. Mishra, B., Meyer, G., Bonnabel, S., Sepulchre, R.: Fixed-rank matrix factorizations and Riemannian low-rank optimization. Comput. Stat. **29**(3–4), 591–621 (2014)
17. Rudelson, M., Vershynin, R.: On sparse reconstruction from Fourier and Gaussian measurements. Commun. Pure Appl. Math. **61**(8), 1025–1045 (2008)
18. Schneider, R.: Convex Bodies: The Brunn-Minkowski Theory 151. Cambridge University, Cambridge (2014)
19. Schneider, R., Weil, W.: Stochastic and Integral Geometry. Springer, Berlin (2008)
20. Tsakiris, M.C., Peng, L., Conca, A., Kneip, L., Shi, Y., Choi, H.: An algebraic-geometric approach to shuffled linear regression. arXiv preprint:1810.05440 (2018)
21. Vershynin, R.: Introduction to the non-asymptotic analysis of random matrices. In: Compressed Sensing, Theory and Applications, pp. 210–268 (2010)

22. Yatawatta, S.: On the interpolation of calibration solutions obtained in radio interferometry. Mon. Not. R. Astron. Soc. **428**(1), 828–833 (2013)
23. Yatawatta, S.: Radio interferometric calibration using a Riemannian manifold. In: 2013 IEEE International Conference on Acoustics, Speech and Signal Processing (ICASSP), pp. 3866–3870. IEEE, Piscataway (2013)

Printed in the United States
by Baker & Taylor Publisher Services